AF805204

Dynamics of Suspended Flow in Open Channels

Aybek M. Arifjanov · Xi Chen ·
A. Fatxulloev Mirzotilloevich ·
Luqmon N. Samiev

Dynamics of Suspended Flow in Open Channels

Springer

Aybek M. Arifjanov
Department of Hydraulics
and Hydroinformatics
"Tashkent Institute of Irrigation
and Agricultural Mechanization Engineers"
National Research University
Tashkent, Uzbekistan

A. Fatxulloev Mirzotilloevich
Department of Hydrology
and Hydrogeology
"Tashkent Institute of Irrigation
and Agricultural Mechanization Engineers"
National Research University
Tashkent, Uzbekistan

Xi Chen
College of Geoinformatics
Zhejiang University of Technology
Hangzhou, China

Luqmon N. Samiev
"Tashkent Institute of Irrigation
and Agricultural Mechanization Engineers"
National Research University
Tashkent, Uzbekistan

ISBN 978-981-96-8780-0 ISBN 978-981-96-8781-7 (eBook)
https://doi.org/10.1007/978-981-96-8781-7

© The Editor(s) (if applicable) and The Author(s), under exclusive license to Springer Nature Singapore Pte Ltd. 2026

This work is subject to copyright. All rights are solely and exclusively licensed by the Publisher, whether the whole or part of the material is concerned, specifically the rights of translation, reprinting, reuse of illustrations, recitation, broadcasting, reproduction on microfilms or in any other physical way, and transmission or information storage and retrieval, electronic adaptation, computer software, or by similar or dissimilar methodology now known or hereafter developed.
The use of general descriptive names, registered names, trademarks, service marks, etc. in this publication does not imply, even in the absence of a specific statement, that such names are exempt from the relevant protective laws and regulations and therefore free for general use.
The publisher, the authors and the editors are safe to assume that the advice and information in this book are believed to be true and accurate at the date of publication. Neither the publisher nor the authors or the editors give a warranty, expressed or implied, with respect to the material contained herein or for any errors or omissions that may have been made. The publisher remains neutral with regard to jurisdictional claims in published maps and institutional affiliations.

Responsible Editor: Prof. K. Sh. Latipov, Doctor of Technical Sciences

Reviewers: Prof. E. J. Makhmudov, Doctor of Technical Sciences; I. Khudjaev, Doctor of Technical Sciences

This Springer imprint is published by the registered company Springer Nature Singapore Pte Ltd.
The registered company address is: 152 Beach Road, #21-01/04 Gateway East, Singapore 189721, Singapore

If disposing of this product, please recycle the paper.

Introduction

Increasing the integrated and highly efficient use of water resources, as well as accelerating scientific and technological advancements in water management construction, is impossible without a comprehensive and detailed study of the transport and regulation of suspended solids by turbulent fluid flows. Such flows occur during the transport of suspended sediment particles in both artificial and natural channels, in seas, during the formation of mudflows on slopes, during the enrichment of minerals, in the hydrotransport of solid materials, in chemical technology, and in various other fields.

In the broader context of studying suspended flows, the transport and regulation of suspended sediment particles in both artificial and natural channels are of particular significance. To ensure the reliability of design and the effective operation of hydraulic engineering and drainage structures, as well as to guarantee water intake, it is essential to investigate the patterns of suspended flow movement. This understanding can then inform the development of methods for regulating river sediment regimes.

It is important to note that, to date, extensive information has been gathered on various sediment parameters, including the size and shape of sediment particles, their granulometric composition, hydraulic size, modes of movement, and the process of particle suspension in fluid flow. However, this information is rarely utilized in theoretical studies and engineering calculations.

Thus, this study provides a cumulative analysis of contemporary theories regarding the patterns that govern sediment movement. It generalizes a substantial amount of experimental and field data and develops advanced computational models based on this information. These models incorporate comprehensive details about the physical nature of the processes and factors that characterize the movement of suspended sediment in channels. Addressing the critical issues of transport, regulation, and utilization of river sediments is essential.

In this monograph, a new model for the movement of suspension-carrying flow in channels is developed and substantiated based on theoretical research. A novel generalization of the mathematical model is presented to determine the hydrodynamic parameters of suspended particles, which govern the interaction between the carrier fluid and solid particles. This model accounts for various modes of movement of suspended particles and encompasses a wide range of variations in the characteristics of solid particles within a fluid flow.

The pattern of formation of the velocity field at various saturations of flow with suspension has been established. Based on this, a new method has been developed to calculate the distribution of the average velocity of both the carrier fluid and solid particles along the depth of the flow.

A mathematical model for the process of weighing and transporting suspended particles in a turbulent flow is proposed. A method has been developed to calculate the concentration field of suspended sediment particles throughout the flow depth, incorporating particle size for the first time by considering the diameter of solid particles. Based on an analysis of field data and computational results, the practical applicability of the proposed calculation methods is evaluated.

Scientifically based methods have been developed to calculate the parameters of suspended flow in channels, ensuring the reliability and quality of the design, construction, and operation of hydraulic engineering and drainage structures.

The implementation of the developed recommendations facilitates the efficient regulation of solid flow regimes in riverbeds. Additionally, by reducing operational and construction costs, these recommendations lead to more economical and reliable design solutions.

About This Book

This book provides an overview of modern state theory regarding turbulent flow in riverbeds. It addresses questions related to turbulent motion in these environments. A new generalization of the mathematical model has been presented for determining the hydrodynamic parameters of suspended particles, as well as the interactions between the liquid and solid carrier particles during various movements of suspended particles.

Considering the principles governing the formation of the velocity field at various saturation levels in slurry flow, this study aims to develop a novel method for calculating the distribution of the average velocity of both the carrier liquid and solid particles throughout the depth and length of the stream.

This book is intended for professionals in hydro techniques, hydraulics, hydrology, and melioration, and it may also be beneficial for doctoral researchers and students in these fields.

Contents

Modern Theories and Methods for Describing the Movement of Suspended Flows

1

1.1 Theoretical Foundations of the Movement of Suspension-Carrying Flows in Channels

For many years, researchers have studied river sediments, accumulating a substantial amount of information that describes a relatively accurate model of the movement of solid particles in turbulent flow.

A suspended flow can be regarded as a heterogeneous mixture, typically consisting of two phases. One phase comprises a discrete set of various types of inclusions in the form of solid particles, while the other phase consists of a continuous liquid medium that surrounds these inclusions.

Mathematical descriptions of real heterogeneous mixtures are more complex than those of single-phase flows for two primary reasons. First, it becomes increasingly challenging to characterize processes within individual phases—such as compressibility, viscosity, and turbulence—that also occur in single-phase media. Second, in multiphase systems, in addition to the aforementioned factors, the effects of phase structure and its variations, as well as interphase interactions (including momentum and energy exchange at the interface, chaotic movements, particle collisions, crushing, and coagulation), are significantly pronounced. Consequently, the range of possible processes that must be represented in the equations expands dramatically. Therefore, it is crucial to develop a unified framework for accounting for several key effects, utilizing theoretical analysis data where possible, and empirical relationships and parameters where necessary.

Kh. A. Rakhmatulin, F. I. Frankl, G. I. Barenblatt, V. M. Makkaveev, O. F. Vasiliev, and M. A. made significant contributions to the theory of constructing mathematical models related to the study of two-phase (multiphase) flows. Additionally, Dementyev, M. A. Velikanov, O. G. Natishivili, R. I. Nigmatulin, D. F. Faizullaev, K. Sh. Latipov, A. I. Umarov, V. M. Lyakhter, Yu. A. Buevich, S. I. Krill, S. Sow, G. Woilis, A. Fortier, and other researchers have also contributed to this field.

© The Author(s), under exclusive license to Springer Nature Singapore Pte Ltd. 2026
A. M. Arifjanov et al., *Dynamics of Suspended Flow in Open Channels*,
https://doi.org/10.1007/978-981-96-8781-7_1

However, the complexity of transporting suspended particles in fluid flow within channels has often hindered the successful application of the aforementioned models to describe suspension-carrying flows.

The most widely used theories in the study of suspension-carrying flows in channels are semi-empirical: diffusion and gravity. These theories are described in detail in the works of numerous scientists.

The distribution of velocity, turbidity, and the size of solid particles along the depth and length of a suspended flow has long captured the attention of researchers. A more precise solution to these issues, even within a one-dimensional framework, could unlock new opportunities for advancing our understanding of the structure of turbulent suspended flows. Below, we will explore the potential applications of existing theories in this context.

V. M. Makkaveev [105], building on the concepts introduced by Taylor and Schmidt (1925), was the first to derive a general differential equation for turbulent diffusion concerning the movement of suspended particles in a water flow. In this context, it is typically assumed that the presence of particles does not disrupt the flow kinematics. This statement holds true only for a small concentration of relatively fine sediment, assuming that the coefficients of turbulent exchange for the transported substances are equal and that the number of movements is consistent.

$$\frac{\mathrm{d}s}{\mathrm{d}t} = \frac{1}{p}\left[\frac{\partial}{\partial x}\left(A\frac{\partial s}{\partial x}\right) + \frac{\partial}{\partial y}\left(A\frac{\partial s}{\partial y}\right) + \frac{\partial}{\partial z}\left(A\frac{\partial s}{\partial z}\right)\right] + \omega\frac{\partial s}{\partial y}, \tag{1.1}$$

where $\frac{\mathrm{d}s}{\mathrm{d}t} = \frac{\partial s}{\partial t} + u\frac{\partial s}{\partial x} + v\frac{\partial s}{\partial y} + w\frac{\partial s}{\partial z}$;

u, v, w—components of the velocity vector in the corresponding directions x, y, z;

s—concentration of suspended particles in water;

x, y, z—coordinate axes of the Cartesian system; the y-axis is directed from top to bottom; the x axis is located on the surface in the direction of the flow; z axis directed normal to x and y;

ρ—flux density;

A—coefficient of turbulent exchange, proportional to the coefficient of turbulent diffusion $\varepsilon = \frac{A}{\rho}$;

ω—hydraulic fineness.

For the case of a flat stationary uniform flow and an established concentration distribution, i.e. $v = w = \frac{d}{s} = 0$; $\frac{\partial s}{\partial z} = \frac{\partial s}{\partial x} = 0$, differential equation (1.1) takes the form

$$\frac{\partial}{\partial y}\left(\varepsilon\frac{\partial s}{\partial y}\right) + w\frac{\partial s}{\partial y} = 0. \tag{1.2}$$

Under the boundary conditions $y = 0$, $s = s_0$, integrating expression (1.2) twice, we obtain

$$s(y) - s_0 \exp\left\{-w\int_0^{y_0}\frac{\mathrm{d}y}{\varepsilon}\right\}. \tag{1.3}$$

Application of Eqs. (1.1) and (1.3) in practice requires determination of the turbulent diffusion coefficient ε-.

There are many proposals based on the use of relation (1.3) for determining the concentration profile of suspended particles in uniform horizontal flows. The resulting formulas differ only due to different proposals about the speed profile $\frac{du}{dy}$. Based on the parabolic velocity profile (Boussinesq, 1876), ε, V. M. Makkaveev proposes the following dependence to determine:

$$\varepsilon = \frac{iyg}{\frac{du}{dy}}, \tag{1.4}$$

where i is the longitudinal slope of the free surface of the flow;

S_0—sediment concentration at the bottom;

u is the average flow velocity.

Expression (1.4) shows that the definition ε depends on the accepted laws of velocity distribution over the depth of the flow, i.e. with different $\frac{du}{dy}$, we obtain different expressions ε and, consequently, different concentration distribution profiles along the flow depth. Further developments in the development of the diffusion theory of turbulence are mainly aimed at determining the value of the turbulent diffusion coefficient.

A major contribution to the development of the diffusion theory of sediment suspension and transport was made by M. A. Velikanov [33], A. D. Girgidov [37], V. K. Debolsky [49], I. F. Karasev [72], A. V. Karaushchev [73], I. A. Kuzmin [82], I. I. Levi [97], K. I. Rossinsky [134], B. A. Fidman [153], I. A. Sherenkov [171], S. I. Krill [78], H. Rouse [126], etc.

However, the application of the theory is often constrained in various instances due to specific assumptions.

In the literature [33, 78, 82, 100], several comments have been made regarding diffusion theory. Firstly, the theory relies on the assumption that the coefficients of turbulent exchange, transferred substances, and momentum are equal. However, at high concentrations, the exchange of substances decreases in comparison to the exchange of momentum. Secondly, the impact of suspended particles on the structure of turbulent flow is not considered. Thirdly, the physical nature of the turbulent exchange coefficient remains unclear.

M. A. Velikanov [33] developed a gravitational theory positing that work must be expended to keep sediment in suspension. He notes that in the case of uniform movement of a homogeneous fluid with a free surface, two mutually balancing forces are at play: gravity, which drives the fluid down the slope, and the resistance force, which opposes this movement.

M. A. Velikanov considers the role of work in the energy balance of averaged motion as follows: for a flow free of sediment, the work done by gravity can be equated to the work done by resistance forces. In the case of a flow that carries sediment, he deems it essential to include the work of weighing in addition to the work of the resistance forces.

Then, the energy balance equation for the liquid phase has the form:

$$\rho giu(i - s) = \rho u \frac{\mathrm{d}}{\mathrm{d}y}\left[(1 - s)\overline{u'v'}\right] + gws(1 - s)(\rho_T - p), \tag{1.5}$$

where $\overline{u'v'}$—parameter of turbulent momentum transfer;

ρ_t is the density of the solid particle.

A similar equation for the solid phase is written as

$$\rho_{\mathrm{T}} gius = \frac{\mathrm{d}}{\mathrm{d}y}\left(\rho_{\mathrm{T}} s\overline{u'v'}\right)u - (\rho_{\mathrm{T}} - \rho)gws(1 - s) \tag{1.6}$$

and obtained the following dependence for calculating the turbidity distribution over the flow depth [35]:

$$s(y) = s_0 \exp\left[-\int_{\eta_0}^{\eta} \frac{\mathrm{d}\eta}{(1 - \eta)\ln\left(1 + \frac{\eta}{\xi}\right)}\right]^{\beta}, \tag{1.7}$$

where $\beta = \frac{a æ w}{i\sqrt{gHi}}$; $a = \frac{\rho_{\mathrm{T}}}{\rho} - 1$;

$\eta = \frac{y}{H}$—relative depth;

ξ—relative roughness;

N—flow depth;

æ—Karman parameter.

A comprehensive presentation of gravitational theory can be found in Refs. [33, 19, 146].

The primary objections to M. A. Velikanov's gravitational theory were articulated by A. N. Kolmogorov. These objections centred on the notion that the energy expended by the flow in transferring suspended particles is derived from pulsation energy rather than directly from the energy of averaged motion. Consequently, it is essential to incorporate this energy into the balance of the pulsating energy of the flow.

M. A. Fidman and others also support this perspective. The most comprehensive experimental analysis of gravitational theory was conducted by N. A. Mikhailova in the book [109], which highlights the discrepancies between this theory and experimental results.

The works of G. I. Barenblatt [2, 6] played a significant role in the development of the theory of suspended flow. He examines flows that transport small particles with a relatively low volume of suspended matter and derives a complex system of equations. To solve the proposed equations, the author makes several simplifying assumptions, resulting in dependencies akin to those found in diffusion theory.

The theory of G. I. Barenblatt was further developed in the works of O. G. Natishvili [4, 11, 115], which incorporate the resulting equations of suspension flow and account for potential slip between phases.

Fundamental contributions to the theory of suspended turbulent flow are attributed to the works of F. I. Frankl. His system of general differential equations [1, 56] encompasses the equations of continuity, momentum, and energy for both averaged and pulsating motion. A notable achievement of F. I. Frankl is the rigorous formulation of averaging operations concerning the motion of a two-component fluid, which led to the derivation of equations devoid of concentration terms. However, the resulting system of equations was found to be open. To close this approximate system, F. I. Frankl proposed several assumptions, suggesting that the fluid flow field could be treated as known and independent of the presence of solid particles. Consequently, he focused on a system of equations that excludes all components of turbulent stresses. Nevertheless, the practical application of even this simplified expression of the equations remains quite challenging [1, 51, 79, 159].

F. I. Frankl's theory was further developed in the works of A. D. Yunin [55], B. A. Fidman [153], S. I. Krill [79], and others.

In this theory, each medium is regarded as continuous, and its movement is viewed as occurring within a porous medium formed by the remaining phases. The interaction between the media is considered, along with the friction between them, which is proportional to their relative velocity.

The system of equations presented in this work is closed, allowing for a theoretical study of various phenomena that occur in a mixture during its movement.

The theory of interpenetrating media [133] was further developed in the works of D. F. Faizullaev [148], R. I. Nigmatullin [116], K. Sh. Latipov [86], I. Umarov [147], and others. K. Sh. Latipov [86, 87] applied this theory to the conditions of channel flows and derived dependencies that were confirmed by experimental and field measurements.

The review indicates that, currently, despite some successes in developing mathematical models for the movement of multiphase (two-phase) flows, the formulation of a comprehensive theory for suspension-carrying flows remains incomplete.

For successful application in practice, existing differential equations must be closed by incorporating straightforward additional dependencies.

1.2 Possibility of Using Existing Methods for Practical Solution of Problems of Suspension Flows

The practical applications of studies on river sediment dynamics are extensive. Designing hydraulic structures and canals necessitates consideration of sediment transport dynamics and the impact of constructed structures on the movement of solid runoff.

The regime of solid runoff, along with its intra-annual and long-term variability and sediment size, constitutes essential initial data for the design of both large and small reservoirs, extensive canals, and settling tanks. This information is crucial for the efficient operation of these structures and for predicting channel deformations.

Many researchers have contributed to the development of practical methods for calculating sediment suspension and transport. Notable among them are the works of S. H. Abalyanets, V. S. Altunin, B. K. Balakaev, O. Brien, V. Vanoni, V. N. Gancharov, V. Graf, A. N. V. Grishanin, V. K. Debolsky, I. V. Egiazarov, E. A. Zamarin, Yu. A. Ibad-Zade, H. A. Irmukhamedov, H. A. Ismagilov, A. A. Kadyrov, A. V. Karausheva, I. F. Karaseva, M. R. Karapetyan, S. I. Krilya, I. A. Kuzmina, K. Sh. Latipova, I. I. Levi, N. A. Mikhailova, M. A. Mikhalev, A. M. Mukhamedova, I. K. Nikitina, E. K. Rabkova, N. A. Rzhanitsyn, K. I. Rossinsky, A. G. Khachatryan, G. Shamov, Kh. Sh. Shapiro, H. A. Einstein, N. Chin, and others.

The development of calculated dependencies for determining the distribution of velocity, turbidity, and particle size along the flow depth plays a crucial role in the study of suspended flow dynamics. It is no coincidence that the capabilities of existing theoretical frameworks have frequently been tested in addressing these challenges.

As mentioned earlier, the solution to these problems is primarily based on the diffusion theory of turbulence. The initial equation for determining the distribution of turbidity along the flow depth is represented by formula (1.3). V. Makkaveev [105] adopted a parabolic vertical velocity distribution profile

$$u = u_{\max}\left(1 - my^2\right), \tag{1.8}$$

where: $u_{\max}$—averaged longitudinal velocity on the surface of the flow;

y—ordinate of a given point, counting from the surface;

m—parameter, accepted within 20–24.

Using relations (1.4) for the distribution of turbidity along the flow depth, V. M. Makkaveev obtains the following dependence:

$$S = S_0 \exp\left\{-\frac{2mC}{g} - \frac{w}{v_{\text{cp}}}(1 - \eta)\right\}, \tag{1.9}$$

where: C—Chezy coefficient;

v—average flow speed.

In this case, the diffusion turbulence coefficient ε taken as a constant value. However, further use revealed variability in the turbulent diffusion coefficient [70, 79, 118].

A. V. Karaushev [73] to determine the coefficient of turbulent exchange takes the formula

$$A = \frac{\rho g H u}{MC} \tag{1.10}$$

and for the speed distribution—the elliptic law

$$u = u_{\max}\sqrt{1 - P\eta^2}, \tag{1.11}$$

where P—dimensionless parameter, which is determined by an approximate formula depending on the Chezy coefficient;

η—relative depth;

M—parameter determined from the expression

$$M = PC\frac{U_{\max}^2}{v_{cp}}. \tag{1.12}$$

After substituting (1.10) and (1.11) into (1.3), the formula for the distribution of partial turbidity s_i by flow depth, expressing the sediment content of the *ith fraction* in water, takes the form:

$$s_i = s_{0i}\exp\{-a\tau_i\beta\rho(y)\}, \tag{1.13}$$

where $\beta(y) = \arcsin\sqrt{p} - \mathrm{arcsom}\left(y\sqrt{p}\right)$; $a = \frac{C\sqrt{M}}{g}$; $\tau_i = \frac{w_i}{v_{cp}}$.

In the case of heterogeneous sediments, the calculation of the distribution of total turbidity is performed by adding partial diagrams (the method was first proposed by A. V. Karaushev):

$$s(y) = \sum_{i=1}^{m_0} s_i(y), \tag{1.14}$$

where: $s_i(y)$—calculated using formula (1.13);

m_0—the number of all sediment fractions suspended in the flow.

The elliptic law of velocity distribution was further developed in the works of A. V. Karaushev's students.

To describe the nature of the vertical distribution of turbidity, A. D. Greshayev adopted an elliptical law of velocity distribution with a free term and obtained another dependence for the distribution of turbidity along the depth of the flow.

Yu. A. Ibadzade [65], having compared the calculation results using A. D. Greshaev's formula, noted a large discrepancy between the calculation results and the measured data.

I. I. Levi [98], taking the distribution of speeds according to the exponential law in the form

$$u = \frac{\sqrt{ghi}}{\lambda}\left[\left(\frac{H}{\Delta} - 1\right)^x + 1 - \left(\frac{y}{\Delta}\right)\right], \tag{1.15}$$

where: λ—coefficient of hydraulic friction;

Δ—height of the roughness protrusion;

x—dimensionless parameter;

to determine the distribution of turbidity along the depth of the flow, I obtained the following relationship:

$$s = s_0\exp\left\{\frac{xw}{1-x}\cdot\frac{u_{\max}}{\lambda v_{cp}^2}\left[1 - \left(\frac{y}{H-\Delta}\right)^{x-1}\right]\right\}. \tag{1.16}$$

It is appropriate to note here that the exponential law of velocity distribution has been satisfactorily used in many cases in solving practical problems [5, 47, 73, 75, 115], since the boundary conditions are successfully formulated in it. However, the lack of a clear theoretical basis limits the use of the formula.

M. V. Velikanov [33, 34], based on the logarithmic law of velocity distribution along the flow depth, obtains the expression for the turbidity distribution:

$$s = s_0 \left[\frac{(1-\eta)}{\xi + \eta} \right]^{\frac{w}{\mathcal{H} C \sqrt{gHi}}}. \tag{1.17}$$

Despite the advantages of the logarithmic law, selecting appropriate boundary conditions can present certain challenges.

The logarithmic law of velocity distribution was also adopted in Refs. [25, 34, 112, 116, 165, 185, 186]. The primary distinction is reflected in the definition of the Kármán parameter, κ.

Research has established that the variability of the parameter æ often depends on the saturation of the flow with suspended matter. This once again underscores the necessity of considering the uneven distribution of turbidity when deriving the equation for the distribution of average velocity throughout the depth of the flow.

An analysis of formulas for the distribution of turbidity along the flow depth, based on diffusion theory, has been conducted by several authors [109, 112, 83, 126, 140, 160]. The results of this analysis yield conflicting conclusions. For instance, A. K. Proskuryakov [121], utilizing Vanoni's experimental data, favours M. A. Velikanov's formula (1.17), as he identifies a significant discrepancy with the results obtained using V. M. Makkaveev's formula (1.19).

Kh. Sh. Shapiro [170], utilizing various experimental data, concludes that the calculations using formula (1.9) yield satisfactory results, while the calculations using formula (1.17) significantly deviate from the experimental data.

V. K. Debolsky [49] compares Formula (1.13) with experiments on Bakelite and points out a significant divergence between (1.13) and the experimental results. In contrast, B. K. Balakaev [25], analysing natural data, observes a satisfactory similarity between formula (1.13) and field data.

This inconsistency is seemingly linked to the accepted values of the parameters (æ) included in the formula for the distribution of velocities along the depth of the flow. The studies conducted by several authors [5, 13, 62, 64, 65, 180, 212, 217, 221] indicate that the parameters exhibit variability within wide limits, which restricts the applicability of the aforementioned formulas.

It is important to note that the laws under consideration do not account for the quantitative influence of sediment on velocity distribution. This factor is significant in the study of suspension flow patterns [120, 128, 100, 144].

M. A. Dementyev and M. Pechenkin [52] propose a formula for determining the distribution of suspended particles in a pipeline.

$$s = s_0 \exp\left[-\frac{A_1}{\theta^2} (\eta - \eta_0) \right], \tag{1.18}$$

where $\theta^2 = C_0 \frac{W^{\frac{1}{2}}}{2}$; $A_1 = K_S\left(\frac{\rho_0 \cdot g}{p} - 1\right)\frac{D_S}{2}$; $\eta = \frac{y}{D_s}$;

w—root mean square value of the pulsation component of the instantaneous relative velocity of a suspended particle;

D_s—height of hydraulic mixture flow;

C_0—drag coefficient of suspended particles;

K_s—experimental coefficient.

It is important to note that scientific knowledge is undoubtedly advantageous in formula (1.18). However, in the case of open flows, determining the accepted initial parameters necessitates specialized research.

A. S. Obrazovsky [1, 18] describes the distribution of turbidity along the depth of the flow based on the turbulent flow movement model proposed by N. M. Bernadsky. The dependence he derived characterizes the general concept of turbidity distribution. This concept was further developed in subsequent works [75, 87].

A number of studies have been conducted to determine the relationship between the transfer coefficient of suspended particles and the momentum transfer coefficient. Various formulas have been developed to characterize the distribution of turbidity with depth in the flow.

V. S. Sinelshchikov [14] proposed a formula for calculating the distribution of turbidity with depth in cases of high concentration.

$$s = s_0 \frac{D_n}{hu \cos \vartheta}\left[1 - \exp\left(\frac{uh \cos \vartheta}{D_n}\right)\right], \tag{1.19}$$

where: $D_n = N_n + \frac{u}{g} b_y$—eddy diffusion coefficient;

ϑ—the angle between the vertical and the flow axis;

b_y—the intensity of the transverse pulsation movement of the stratified liquid;

N_n—eddy viscosity coefficient.

Definition of b_y is difficult, since existing measuring instruments cannot measure its value with sufficient accuracy.

S. I. Kril [78] obtained the formula for determining the distribution of turbidity along the depth of the pipeline by summarizing the developments of diffusion and gravitational theories, based on the Barenblatt gravitational model.

$$\varepsilon_s \frac{\mathrm{d}s}{\mathrm{d}\eta} = \frac{Y_{\text{kp}}}{Y_{\text{cp}}} C_1 (1 - C_1)^{\overline{\Delta}}, \tag{1.20}$$

where: $C = \frac{s}{s_m}$—ratio of concentration s to the maximum possible suspended matter concentrations s_m;

η—relative depth;

Y_{kp}—critical speed of hydrotransport;

$\overline{\Delta}$—exponent depends on Reynolds number $\mathrm{Re_d}$.

Due to the differing nature of the movement of suspended particles in channels and pipes, the application of S. I. Krill's formula presents several challenges. Firstly, when suspended particles flow through pipes, the flow does not become saturated at the bottom. Secondly, determining the critical speed in channels necessitates considerable effort.

Taking into account the randomness and chaotic nature of suspended particle movement in a flow, some researchers have employed probability theory methods to describe the behaviour of solid particles [33, 38, 106, 25, 131, 190].

N. V. Bernatskaya [2, 9] derives the equation for the distribution of turbidity and the size of suspended particles along the depth of the flow by utilizing probability theory.

When calculating the distribution of turbidity and particle size using the method outlined in [8], the initial data consists of the average particle size of the bottom material and the dispersion of its granulometric composition. However, studies indicate that not all particles transported by the flow can be observed at the bottom [1, 4, 5], and often a significant number of suspended particles are carried by the flow in transit.

The importance of considering the particle size distribution throughout the flow depth when deriving calculated dependencies has been emphasized in the works of several researchers [20, 79, 88, 112, 140, 160].

M. A. Mamedov [108] proposes a formula to describe the size distribution in depth

$$d_y = \frac{b_1 (u + v)^2 w_{\text{cp}}}{gy\left(\eta + \frac{w}{u} + a_1\right)},$$

where d_y—particle size at a distance y from the bottom;

a_1, b_1—dimensionless coefficients;

w_{cp}—medium hydraulic size.

Analysing the formula proposed by M. A. Mamedov, V. K. Debolsky [49], and N. V. Bernatskaya [29] indicates a divergence between the calculated results and the measured data.

Thus, based on the analysis conducted, it is evident that the potential for successfully applying the examined formulas for calculating the distribution of average velocities, turbidity, and particle size along the depth of a suspended open flow is limited. The pursuit of new relationships that are user-friendly and account for various processes inherent in suspension flows will be the focus of future research.

To date, numerous studies have demonstrated the significance and necessity of investigating the movement of suspended particles within a flow.

The work [136, 119, 82] emphasizes the importance of considering suspended particles not only when addressing individual problems but also in describing the principles of channel formation.

As noted above (1.1), one of the primary methods for studying turbulent suspension-carrying flows is to develop a closed system of differential equations that comprehensively describe the characteristics of turbulent two-phase flows, or to identify new dependencies that can be used to close existing equations.

Many rivers, such as the Amu Darya, and large canals transport a significant quantity of fine solid particles, resulting in a liquid that differs from pure water in terms of density and viscosity.

I. I. Levi and N. P. Kulesh [98] concluded that carried by the flow should be divided into two groups: particles smaller than 0.015 mm and particles larger than 0.015 mm. The former forms a suspension with water—a homogeneous liquid that serves as a medium for carrying larger sediments, which are mechanical impurities. Therefore, it is essential to consider various factors that characterize the movement of suspension-carrying flow in channels for a more comprehensive understanding of the process.

This paper presents a generalization of contemporary concepts, identifying a pattern that describes the movement of suspended flow in channels. Based on this pattern, a calculation model is developed to address the scientific and practical challenges associated with the transport, regulation, and utilization of river sediment.

Mathematical Model of the Movement of Suspended Flow in Open Channels

2

2.1 Basic Equations of Motion of Suspension-Carrying Flow in Channels

As noted above, many researchers have contributed to the development of a mathematical model for the movement of two-phase (multiphase) flows. When applying these models to the movement of suspended flows in channels, the following conditions must be adhered to:

- Align the movement model more closely with the physical principles governing suspended flow.

The closed nature of the system of differential equations must be maintained.

When a suspended flow is treated as a single-phase continuum—meaning that differential equations are formulated solely for the liquid phase—challenges arise in analyzing the flow structure, particularly in determining the kinematic characteristics of both the carrying liquid and the solid particles. Therefore, to effectively describe and simulate the dynamics of suspended flows, it is currently more advantageous to adopt a model of conditionally continuous, interpenetrating, and interacting media, each defined by its own governing equations. In this framework, the suspended flow is regarded as a two-phase system, where each phase of the mixture possesses distinct physical properties (such as density, viscosity, and velocity). This approach allows for the separate examination of the parameters associated with each phase.

The theory of motion for interpenetrating multiphase (multicomponent) mixtures, formulated by Kh. A. Rakhmatulin [133], has been extensively applied to address practical problems in this field. The majority of the problems solved using this theory have involved examples of laminar flow. Subsequent research in this area has demonstrated that the applicability of the theory, under certain conditions, extends well beyond the confines of laminar motion.

© The Author(s), under exclusive license to Springer Nature Singapore Pte Ltd. 2026

A. M. Arifjanov et al., *Dynamics of Suspended Flow in Open Channels*,
https://doi.org/10.1007/978-981-96-8781-7_2

When formulating the equation of motion, the suspension-carrying flow is regarded as a hydraulic mixture composed of two phases: a liquid carrier and solid particles, with the liquid phase serving as the primary carrier.

When analyzing the movement of each phase individually, external forces are applied to each phase. Additionally, a force develops between the phases, which, when considering the mixture as a whole, is classified as an internal force. Consequently, the forces acting on each phase comprise surface forces, mass forces, and interaction forces. Therefore, the equations of motion for the *i*th phase can be expressed in differential form as follows:

$$\rho_1 \overline{a}_1 = \operatorname{div} \overline{P}_i + \overline{\Pi}_i + \rho_i \overline{F}_i, \tag{2.1}$$

here $P_{ij} = -f_i \operatorname{grad} P + \frac{\partial \tau_{ij}}{\partial x_{ij}}$—stress tensor acting on the *i*-phase;

τ_{ij}—internal stresses;

f_i—volumetric content of the *i* phase ($i = 1, 2$).

In the case of laminar flow, this tensor is equivalent to the viscous stress tensor [100, 120, 133, 156].

In the case of turbulent motion, it comprises the sum of viscous stress and Reynolds stress, generalized for a multiphase mixture [128]. The contributions of A. N. Frankl, B. A. Fidman, A. K. Dyunin, and others are well-documented in this area.

According to the aforementioned studies, the components of the generalized Reynolds stress for a multiphase fluid are functions of averaged velocities, concentrations, and the first and second derivatives of averaged velocities. However, it is well known that when addressing practical problems using this approach, several mathematical challenges arise. The resulting systems of equations are not closed for both the suspended particles and the carrier fluid, and they cannot be closed in a definitive manner due to the presence of a significant number of additional unknowns.

The P_i components of the interaction force express the influence of an arbitrarily different phase on the *i*th phase. A characteristic feature of this force is that it acts as an external force for the *i*th phase while being internal to the mixture flow as a whole.

It is important to recognize that accounting for the interaction forces in suspension-carrying flows is a crucial aspect when analyzing the flow of carrier liquids and solid particles separately [23, 50, 84, 95, 99, 120, 162]. Each phase within the moving volume possesses its own velocity at any given point, which leads to mechanical interactions between the phases.

It comprises two components:

$$P_i = P_{i1} + P_{i2} \tag{2.2}$$

P_{i1}—force arising due to viscous stresses between individual phases; in a suspended flow, taking into account the insignificance of viscous stresses, the influence of P_2 can be neglected;

P_{i2} is the force caused by the difference in the velocities of the liquid phase at two opposite points of the particle when it consists of phases of particles of solid material.

The forces P_{i1} and P_{i2} characterize discreteness and eliminate the error that arose during the transition to averaged parameters.

Based on the above, it can be noted that the force of interaction in the most general concept is a function of concentrations, velocities and their derivatives, particle size, etc.

$$\overline{\Pi_i} = f\left(U, U_s, f_i\frac{\partial U}{\partial x_j}, \frac{\partial U_s}{\partial x_j}, d, \ldots\right)$$

According to the works [43, 45, 84, 82, 99], the interaction force depending on the speed difference of each phase is considered in the form:

$$\overline{\Pi_i} = K(\overline{U} - \overline{U_8}) \tag{2.3}$$

where $\overline{U}, \overline{U_s}$ are the velocity vectors of the carrier fluid and solid particle, respectively;

K—coefficient of interaction between phases (carrying liquid and solid particle). A detailed analysis of the expression for the interaction force is discussed below (Chap. 3);

F_i—mass forces, in this case the components of gravity are considered mainly.

After considering the main forces characterizing the movement of the suspension-carrying flow, we write Eq. (2.3) in the Cartesian coordinate system in the form:

$$\rho_i\frac{\mathrm{d}u_i}{\mathrm{d}t} = \operatorname{div} P_k + K(U - U_8) + \rho_i F_i.$$

After some simplifications, using the stress tensor expressions, we have:

$$\rho_i\frac{\mathrm{d}u_i}{\mathrm{d}t} = -f_k \operatorname{grad} P + \frac{\mathrm{d}\tau_{kj}}{\mathrm{d}x_j} + K(U - U_s) + \rho_i F_i. \tag{2.4}$$

The value of internal stresses is determined based on work [91, 92, 94, 98].

$$\tau_{ijl} = f_i\mu_i\left(-\frac{2}{3}\operatorname{div}\overline{V_k} + 2\frac{\partial v_{kj}}{\partial x_j}\right) - \int LV_i \mathrm{d}n, \quad npu\ j = l\ (\overline{n} = 1, 2, 3);$$

$$\tau_{ijl} = f_i\mu_i\left(\frac{\partial v_{kl}}{\partial x_j} + \frac{\partial v_{kj}}{\partial x_j}\right), \quad npu\ j \neq l. \tag{2.5}$$

It should be noted that here the parameter L is probabilistic in nature, satisfying the Gaussian distribution, L. The mass transfer between phase layers and for a single-phase flow is determined by the expression [91, 92]:

$$L = \frac{La\mu}{l^2}; \quad La^2 = 0.0025\left(\frac{1-a_1}{1+b_1}\right)\mathrm{Re}(1-\varphi_1) \tag{2.6}$$

where l—dimensional characteristic of the flow;

Re—Reynolds number;

a_1, b_1—coefficients depending on the roughness of the channel:

$$a_1 = \frac{\Delta}{l}; \quad b_1 = \left(\frac{\Delta}{l}\right)^{0,2};$$

φ_1—probability function satisfying the Gaussian distribution.

Bearing in mind that $j = x, y, z$; $l = x, y, z$; u, v, w the velocity components $\overline{U_i}$, summing Eq. (2.5) along the coordinate axis over the same indices, we write it in the form e (for the first phase):

$$\begin{aligned} \rho_1\frac{\mathrm{d}u_1}{\mathrm{d}t} &= -f_k\frac{\partial P}{\partial x} + \frac{\partial \tau_{1xx}}{\partial x} + \frac{\partial \tau_{1xy}}{\partial y} + + \frac{\partial \tau_{1xz}}{\partial z} - K(u_1 - u_2) + \rho_1 F_x; \\ \rho_1\frac{\mathrm{d}v_1}{\mathrm{d}t} &= -f_k\frac{\partial P}{\partial y} + \frac{\partial \tau_{1yx}}{\partial x} + \frac{\partial \tau_{1yy}}{\partial y} + \frac{\partial \tau_{1yz}}{\partial z} - K(v_1 - v_2) + \rho_1 F_y \\ \rho_1\frac{\mathrm{d}w_1}{\mathrm{d}t} &= -f_k\frac{\partial P}{\partial z} + \frac{\partial \tau_{zxx}}{\partial x} + \frac{\partial \tau_{1zy}}{\partial y} + \frac{\partial \tau_{1zz}}{\partial z} - K(\tilde{\omega}_1 - \tilde{\omega}_2) + \rho_1 F_z; \end{aligned} \tag{2.7}$$

where $\frac{du_1}{dt} = \frac{\partial u_1}{\partial t} + u\frac{\partial u_1}{\partial x} + v\frac{\partial u_1}{\partial y} + \tilde{\omega}\frac{\partial u_1}{\partial z}$.

Using expressions (2.5), we write the tangential voltage of the phases in the form:

$$\begin{aligned} \tau_{1xx} &= f_1\mu_1\left(-\frac{2}{3}\left(\frac{\partial u_1}{\partial x} + \frac{\partial v_1}{\partial y} + \frac{\partial \tilde{\omega}_1}{\partial z}\right) + 2\frac{\partial u_1}{\partial x}\right) - \int Lu\,\mathrm{d}x \\ \tau_{1xy} &= f_1\mu_1\left(\frac{\partial u_1}{\partial y} + \frac{\partial v_1}{\partial x}\right) \\ \tau_{1xx} &= f_1\mu_1\left(\frac{\partial u_1}{\partial z} + \frac{\partial \tilde{\omega}_1}{\partial x}\right) \\ \tau_{1xy} &= \tau_{1yx} \\ \tau_{1yy} &= f_1\mu_1\left(-\frac{2}{3}\left(\frac{\partial u_1}{\partial x} + \frac{\partial v_1}{\partial y} + \frac{\partial \tilde{\omega}_1}{\partial z}\right) + 2\frac{\partial v_1}{\partial y}\right) - \int Lv\,\mathrm{d}y \end{aligned} \tag{2.8}$$

$$\tau_{1yz} = f_1\mu_1\left(\frac{\partial v_1}{\partial z} + \frac{\partial \tilde{\omega}_1}{\partial y}\right) \tag{2.9}$$

$$\tau_{1zz} = f_1\mu_1\left(-\frac{2}{3}\left(\frac{\partial u_1}{\partial x} + \frac{\partial v_1}{\partial y} + \frac{\partial \tilde{\omega}_1}{\partial z}\right) + 2\frac{\partial \tilde{\omega}_1}{\partial z}\right) - \int Lw\,\mathrm{d}z$$

$$\begin{aligned}\tau_{1zx} &= \tau_{1xz}\\ \tau_{1zy} &= \tau_{1yz}\end{aligned} \tag{2.10}$$

Substituting systems of Eqs. (2.8)–(2.10) into (2.7), keeping in mind that from the continuity equation div $V_i = 0$ and for an incompressible fluid ($\rho = \mathrm{const}$):

$$\frac{\partial u_1}{\partial x} + \frac{\partial v_1}{\partial y} + \frac{\partial \tilde{\omega}_1}{\partial z} = 0.$$

Then we write relation (2.7) for the first phase of a free-flowing suspension-carrying flow in the form

$$\begin{aligned}
\rho_1\frac{\mathrm{d}u_1}{\mathrm{d}t} &= \frac{\partial}{\partial x}\left(2f_1\mu_1\frac{\partial u_1}{\partial x}\right) + \frac{\partial}{\partial y}\left(f_1\mu_1\left(\frac{\partial u_1}{\partial y} + \frac{\partial v_1}{\partial x}\right)\right)\\
&\quad + \frac{\partial}{\partial z}\left(f_1\mu_1\left(\frac{\partial u_1}{\partial z} + \frac{\partial \tilde{\omega}_1}{\partial x}\right)\right) - Lu_1 + K(u_1 - u_2) + \rho_1F_x\\
\rho_1\frac{\mathrm{d}v_1}{\mathrm{d}t} &= \frac{\partial}{\partial y}\left(2f_1\mu_1\frac{\partial v}{\partial y}\right) + \frac{\partial}{\partial x}\left(f_1\mu_1\left(\frac{\partial u_1}{\partial x} + \frac{\partial v_1}{\partial y}\right)\right)\\
&\quad + \frac{\partial}{\partial z}\left(f_1\mu_1\left(\frac{\partial v_1}{\partial z} + \frac{\partial \tilde{\omega}_1}{\partial y}\right)\right) - Lv_1 + K(v_1 - v_2) + \rho_1F_y\\
\rho_1\frac{\mathrm{d}w_1}{\mathrm{d}t} &= \frac{\partial}{\partial z}\left(2f_1\mu_1\frac{\partial w}{\partial z}\right) + \frac{\partial}{\partial x}\left(f_1\mu_1\left(\frac{\partial w_1}{\partial x} + \frac{\partial u_1}{\partial z}\right)\right)\\
&\quad + \frac{\partial}{\partial y}\left(f_1\mu_1\left(\frac{\partial v_1}{\partial z} + \frac{\partial \tilde{\omega}_1}{\partial y}\right)\right) - Lw_1 + K(\tilde{\omega}_1 - \tilde{\omega}_2) + \rho_1F_z.
\end{aligned} \tag{2.11}$$

For the second phase, accordingly it will be

$$\begin{aligned}
\rho_2\frac{\mathrm{d}u_2}{\mathrm{d}t} &= \frac{\partial}{\partial x}\left(2f_2\mu_2\frac{\partial u_2}{\partial x}\right) + \frac{\partial}{\partial y}\left(f_2\mu_2\left(\frac{\partial u_2}{\partial y} + \frac{\partial v_2}{\partial x}\right)\right)\\
&\quad + \frac{\partial}{\partial z}\left(f_2\mu_2\left(\frac{\partial u_2}{\partial z} + \frac{\partial \tilde{\omega}_2}{\partial x}\right)\right) - Lu_2 + K(u_1 - u_2) + \rho_2F_x\\
\rho_2\frac{\mathrm{d}v_2}{\mathrm{d}t} &= \frac{\partial}{\partial y}\left(2f_2\mu_2\frac{\partial v}{\partial y}\right) + \frac{\partial}{\partial x}\left(f_2\mu_2\left(\frac{\partial u_2}{\partial x} + \frac{\partial v_2}{\partial y}\right)\right)\\
&\quad + \frac{\partial}{\partial z}\left(f_2\mu_2\left(\frac{\partial v_2}{\partial z} + \frac{\partial \tilde{\omega}_2}{\partial y}\right)\right) - Lv_2 + K(v_1 - v_2) + \rho_2F_y
\end{aligned}$$

$$\rho_2\frac{dw_2}{dt} = \frac{\partial}{\partial z}\left(2f_2\mu_2\frac{\partial w}{\partial z}\right) + \frac{\partial}{\partial x}\left(f_2\mu_2\left(\frac{\partial w_2}{\partial x} + \frac{\partial u_2}{\partial z}\right)\right)$$
$$+ \frac{\partial}{\partial y}\left(f_2\mu_2\left(\frac{\partial v_2}{\partial z} + \frac{\partial \tilde{\omega}_2}{\partial y}\right)\right) - Lw_2 + K(\tilde{\omega}_1 - \tilde{\omega}_2) + \rho_2 F_z \quad (2.12)$$

In the case of a one-dimensional flow, the equations of motion of a suspended flow in open channels

$$\frac{\partial u}{\partial x} = \frac{\partial u}{\partial z} = 0$$

$f = s$—concentration of the second phase in the flow.

Following the above, we write Eq. (2.12) in the form:

For suspended particle flow

$$\rho\frac{du_s}{dt} = \frac{\partial}{\partial y}\left(s\mu\frac{\partial u_s}{\partial y}\right) - L_S U_S + K(u - u_S) + s\rho_m g i, \quad (2.13)$$

for carrier fluid flow

$$\rho_m\frac{du}{dt} = \frac{\partial}{\partial y}\left((1 - s)\mu\frac{\partial u}{\partial y}\right) - Lu + K(u_s - u) + (1 - s)\rho g i$$

where ρ, ρ_m—density of the carrier liquid and solid particle;

i—slope of the water surface.

Index s hereinafter indicates the parameters of the flow of suspended particles.

System of Eq. (2.13) describes the movements of a suspension-carrying flow in open channels in a one-dimensional formulation of the problem, which takes into account the forces of interaction of phases, as well as parameters describing the turbulent characteristics (L, L_s) of the suspension-carrying flow and the uneven distribution of suspended particles along the depth of the flow.

2.2 Equation of Motion for Heterogeneous Fractional Composition of Suspended Particles

In cases where the suspended particles are homogeneous, the aforementioned equations are sufficient to describe the process. However, when the particles exhibit heterogeneity in composition, it becomes necessary to formulate equations that account for this variability. To achieve this, we proceed as follows (based on the works of [93, 153]): we construct a curve that illustrates the variation in the content of a fraction along its diameter. The content of the fraction is plotted on the vertical axis, while its diameter is represented on the horizontal axis. An example of such sediment distribution curves in open channels can be found in the works of A. Gostunsky, A. G. Khachatryan, and others.

By approximating this curve through a functional dependence, we obtain:

$$\frac{\partial f_2}{\partial r} = \varphi(r), \quad r_1 \leq r \leq r_2 \tag{2.14}$$

where r_1, r_2—upper and lower limits for changing the radii of sediment particles; $\varphi(r)$—concentration of size fraction *r*.

Then the total content of the fractional composition of suspended particles is determined as follows:

$$f_2 = \int_{r_1}^{r_2} \varphi(r)\mathrm{d}r = \int_{r_1}^{r_2} \frac{\partial f_2}{\partial r}\mathrm{d}r. \tag{2.15}$$

From previous works [93, 133, 153] it is known that for a unit volume the following conditions are satisfied:

$$f_1 + \int_{r_1}^{r_2} \varphi(r)\mathrm{d}r = 1.$$

Then, based on (2.14), (2.15) for any fraction of suspended particles we have a formula of the form:

$$\rho\frac{\partial f_2}{\partial r}\frac{\mathrm{d}u_k}{\mathrm{d}t} = -\frac{\partial f_2}{\partial r}\overline{\mathrm{grad}\,P} + \frac{\mathrm{d}\tau_{kj}}{\mathrm{d}x_1} + \int_{r_1}^{r_2} K(u_l - u_k)\mathrm{d}r + \rho_i F_i, \tag{2.16}$$

for liquid flow

$$\rho\frac{\mathrm{d}u_k}{\mathrm{d}t} = -f_1 \mathrm{grad}\,\overline{P} + \frac{\mathrm{d}\tau_{kj}}{\mathrm{d}x_1} + \int_{r_1}^{r_2} K(u_k - u_l)\mathrm{d}r + \rho_i F_i. \tag{2.17}$$

Then the equation for constant flow will be written as:

For liquid flow

$$\frac{\partial p}{\partial} + \mathrm{div}(p_i v_i) = 0, \tag{2.18}$$

for suspended particle flow

$$\frac{\partial}{\partial t}\left(\frac{\partial f_2}{\partial r}\right) + \mathrm{div}\left(\frac{\partial f_2}{\partial r}\overline{v_i}\right) = 0. \tag{2.19}$$

Equations (2.12), (2.13), (2.16), and (2.17) represent the averaged equations of motion for a suspended flow, applicable to both homogeneous and non-uniform compositions of the suspended particle fraction. Consequently, the general equation governing suspension-carrying flow in channels has been formulated. The next objective is to identify and elucidate the key parameters incorporated in the averaged equations of motion for suspension-carrying flow.

Based on the analysis of existing theoretical developments related to the description of suspension-carrying flow, it has been determined that the complexity of applying these theories to channel suspension-carrying flows arises from the need to account for turbulent characteristics and the closure of the system of differential equations.

Study of Physical and Mechanical Parameters of Suspensive Flow

3

3.1 Movement of a Solid Particle in a Fluid Flow

Studying the characteristics of the turbulent movement of suspended flows is one of the complex problems of engineering hydrology. Its complexity lies in determining the values of various factors that control the movement of two-phase flows, consisting of a carrier liquid and a solid particle.

One of these parameters is the resistance force to the movement of solid particles in a fluid flow, which serves as a prerequisite for establishing numerical values of the main characteristics of suspended sediment particles.

The drag force depends on the nature of the liquid medium flowing around solid particles and the mode of its motion (turbulent or laminar).

A more precise determination of the resistance force presents certain difficulties. As noted in [155], to fully take into account all the factors characterizing the motion of a solid particle in a flow, it is necessary to determine about 30 coefficients. Some scientists have made attempts to establish individual factors characterizing the strength of resistance [31, 39, 45, 47, 49, 59, 61, 68, 117, 124, 137, 159, 185, 189].

The first theoretical solution to this problem was given by Stokes. The Stokes formula was obtained under the condition that the Reynolds number is considered to be very small ($Re \leftarrow 1$); in the equations of motion of a viscous fluid, the quadratic terms of inertia are neglected. It is appropriate to remember here that, despite the limited application of the Stokes formula, it is widely used in colloid chemistry, molecular physics, meteorology, and also in some studies of river sediments.

Further theoretical research in this direction was carried out mainly with the aim of clarifying and testing the Stokes formula (Oseen, Holstein, etc.). However, a more reliable equation for determining the drag force, which would characterize the motion of solid particles even with further changes in Reynolds numbers, has not yet been obtained. Therefore, an approximate solution of the equation of

© The Author(s), under exclusive license to Springer Nature Singapore Pte Ltd. 2026
A. M. Arifjanov et al., *Dynamics of Suspended Flow in Open Channels*,
https://doi.org/10.1007/978-981-96-8781-7_3

motion of a viscous incompressible fluid using the Stokes method remains the most promising.

Following what has been said, let us consider the motion of a particle reduced to the shape of a sphere in a liquid unlimited by water. Based on the works [87,141], we write the equations of motion (2.9), (2.13) in spherical coordinates in the form:

$$\frac{\partial}{\partial R}\left(\frac{1}{2}v^2\right) - \frac{D\varphi}{R^2\sin^2(\theta)}\frac{\partial\varphi}{\partial R} + \frac{1}{\rho}\frac{\partial p}{\partial R} = \frac{D\varphi}{R^2\sin^2(\theta)}\frac{\partial\varphi}{\partial\theta} - \frac{L_s}{R^2\sin^2(\theta)}\frac{\partial\varphi}{\partial\theta}$$
$$\frac{\partial}{\partial R}\left(\frac{1}{2}v^2\right) - \frac{D\varphi}{R^2\sin^2(\theta)}\frac{\partial\varphi}{\partial\theta} + \frac{1}{\rho}\frac{\partial p}{\partial\theta} = -\frac{v}{\sin^2(\theta)}\frac{\partial D\varphi}{\partial R} + \frac{L_s}{\sin(\theta)}\frac{\partial\varphi}{\partial R} \tag{3.1}$$

where φ—current functions;
R, θ—spherical coordinate axes;
v—solid particle speed.

Excluding from Eq. (3.1) using cross differentiation of pressure, we obtain the following differential equations for the stream function:

$$\frac{1}{R^2\sin^2(\theta)}\left(\frac{\partial\varphi}{\partial R}\frac{\partial D\varphi}{\partial\theta} - \frac{\partial\varphi}{\partial R}\frac{2D\varphi}{R^2\sin^2(\theta)}\left(\frac{\partial\varphi}{\partial R}\cos\theta - \frac{\sin\theta}{R}\frac{\partial\varphi}{\partial R}\right)\right)$$
$$= -vDD\varphi + \frac{L_s}{\rho}D\varphi \tag{3.2}$$

where D—Stokes operator:

$$D = \frac{\partial^2}{\partial R^2} + \frac{\sin\theta}{R^2}\frac{\partial}{\partial\theta}\left(\frac{1}{\sin\theta}\frac{\partial}{\partial\theta}\right).$$

With rectilinear translational motion of a ball in an unlimited viscous fluid with a constant speed v, parallel to the axis x (Fig. 3.1), from (3.2) we obtain

$$\mu DD\varphi = L_s D\varphi$$

or

$$DD\varphi = kD\varphi, \tag{3.3}$$

where $k = L_s/\mu$;
μ—dynamic viscosity coefficient.

Pressure and velocity vector projections are determined based on [147]:

$$\frac{\partial p}{\partial R} = \frac{\mu}{R^2\sin(\theta)}\frac{\partial\varphi}{\partial\theta}; \quad v_R = \frac{1}{R^2\sin(\theta)}\frac{\partial\varphi}{\partial\theta}$$
$$\frac{\partial p}{\partial\theta} = \frac{\mu}{\sin(\theta)}\frac{\partial\varphi}{\partial R}; \quad v_\theta = \frac{1}{R\sin(\theta)}\frac{\partial\varphi}{\partial R} \tag{3.4}$$

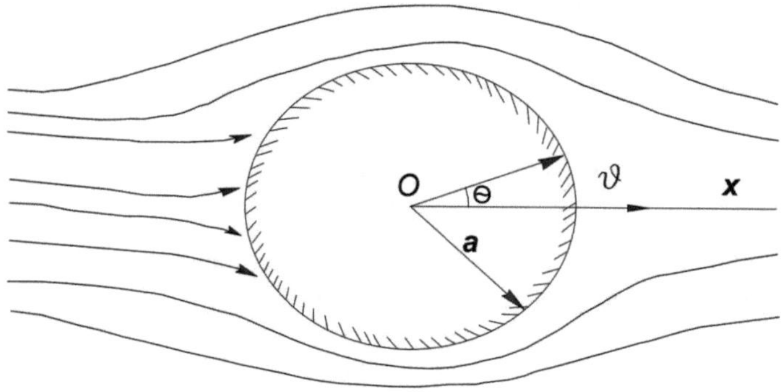

Fig. 3.1 Scheme for calculating the motion of a ball in an unbounded fluid

When solving the problem of translational motion of a ball, we accept the conditions of adhesion to the surface [141]

$$v_R = \frac{1}{R^2 \sin(\theta)} \frac{\partial \varphi}{\partial \theta} = v \cos \theta$$

At $R = a$

$$v_\theta = -\frac{1}{R \sin(\theta)} \frac{\partial \varphi}{\partial R} = -v \sin \theta \tag{3.5}$$

where v_R, v_θ—velocity components in spherical coordinates;
R—radius of the ball.
At

$$R \to \infty, v_R \to 0, v_\theta \to 0. \tag{3.6}$$

The form of boundary conditions (3.5) and (3.6) indicates the possibility of expression for the stream function

$$\varphi = \sin^2 \theta F(R). \tag{3.7}$$

Taking (3.3) into account and substituting (3.7) into (3.3), we obtain

$$D\left[\sin^2 \theta \left(F'' - \frac{2}{R^2} F\right)\right] = k \sin^2 \theta \left(F'' - \frac{2}{R^2} F\right). \tag{3.8}$$

By entering the designation

$$f(R) = F'' - \frac{2}{R^2}, \tag{3.9}$$

from (3.8) we obtain a differential equation for the function f :

$$D\left[\sin^2\theta\left(f - \frac{2}{R^2}f\right)\right] = k\sin^2\theta f(R)$$
$$f'' - f\left(\frac{2}{R^2} + k\right) = 0. \tag{3.10}$$

A particular solution of this (3.10) differential equation is the Bessel function of the imaginary argument of fractional order [145]

$$I_{\frac{3}{2}}(z);\ K_{\frac{3}{2}}(z). \tag{3.11}$$

The general solution is:

$$f = C_1 I_{\frac{3}{2}}(z) + C_2 K_{3/2}(z). \tag{3.12}$$

Substituting (3.12) into (3.10), we obtain a solution to Eq. (3.10) in the form

$$f(R) = R^{1/2}\left[C_1 I_{\frac{3}{2}}(\beta R) + C_2 K_{\frac{2}{3}}(\beta R)\right] \tag{3.13}$$

where is $C_1, C_2 - n$ the constant of integration;

$$\beta = \left(\frac{L_s}{\mu}\right)^{\frac{1}{2}}$$

Using (3.13) from equality (3.9) we have:

$$F'' - \frac{2}{R^2}F = R^{\frac{1}{2}}\left[C_1 I_{\frac{3}{2}}(\beta R) + C_2 K_{3/2}(\beta R)\right] \tag{3.14}$$

We write the solution to the differential equation (3.14) in the form

$$F = A(R)R^2 + \frac{B(R)}{R}. \tag{3.15}$$

Differentiating (3.15) and determining the values of parameters A, B, we obtain a solution to Eq. (3.14)

$$F' = 2A(R) - \frac{B}{R^2} + A'B^2 + \frac{B'}{R\prime}$$
$$F'' = 2A - \frac{B}{R^3} + 2A'R - \frac{B'}{R^2}. \tag{3.16}$$

Substituting expression (3.16) into (3.14), we have:

$$F(R) = R^{\frac{3}{2}}\left[C_1 I_{\frac{3}{2}}(\beta R) - C_2 K_{\frac{3}{2}}(\beta R)\right]/3\beta - R^{\frac{3}{2}}\left[C_1 I_{\frac{3}{2}}(\beta R) - C_2 K_{\frac{3}{2}}(\beta R)\right]/3\beta$$
$$+C_3 R^2 + C_4/R \quad (3.17)$$

where C_3, C_4—integration constants.

Having grouped Eq. (3.17), we can write them in the form

$$F(R) = \frac{C_1 R^{3/2}}{3\beta}\left(I_{3/2}(\beta R) - K_{3/2}(\beta R)\right) + \left[C_3 + \frac{C_1 R^{3/2}}{3\beta} I_{1/2}(\beta R)\right]R^2$$
$$+\frac{\left[C_{43} + \frac{C_1 R^{\frac{3}{2}}}{3\beta} K_{\frac{1}{2}}(\beta R)\right]}{R}. \quad (3.18)$$

Using (3.17) and (3.18), we obtain formulas for the stream function and velocity components

$$\varphi = \left\{\frac{C_1 R^{3/2}}{3\beta}\left(I_{1/2}(\beta R) - I_{5/2}(\beta R)\right) + \frac{C_1 R^{3/2}}{3R}\left[K_{5/2}(\beta R)\right.\right.$$
$$\left.\left. -K_{1/2}(\beta R) + C_5 R^2 + C_4/R\right]\right\}\sin^2(\theta) \quad (3.19)$$

$$v_R = 2\cos\theta\left\{\frac{C_1}{3(\beta R)^{1/2}}\left(I_{1/2}(\beta R) - I_{5/2}(\beta R)\right) + \frac{C_1}{3(\beta R)^{1/2}}\left[K_{5/2}(\beta R)\right.\right.$$
$$\left.\left. -K_{1/2}(\beta R) + C_3 + C_4/R^3\right]\right\} \qquad v_\theta = \sin\theta \quad (3.20)$$

$$\left\{-\frac{2}{3(\beta R)^{1/2}}\left(C_1 I_{1/2}(\beta R) - C_2 K_{1/2}(\beta R)\right) - \frac{2}{3(\beta R)^{1/2}}\left[C_1 I_{1/2}(\beta R)\right.\right.$$
$$\left.\left. -C_2 K_{5/2}(\beta R) - 2C_3 + C_4/R^3\right]\right\}. \quad (3.21)$$

From the boundary conditions it follows that to satisfy conditions (3.6) it is necessary $C_1 = 0$; $C_3 = 0$.

Using boundary conditions (3.5), we obtain a system of equations for determining the constant integration of C_2, C_4.

$$\frac{2C_2}{2\beta(a)^{1/2}}\left(K_{3/2}(\beta a) - K_{1/2}(\beta a)\right) + \frac{2C_4}{a^3} = v$$

$$\frac{C_2}{2\beta(a)^{1/2}}\left(K_{3/2}(\beta a) - 2K_{1/2}(\beta a)\right) + \frac{C_4}{a^3} = v$$

From this system of equations we have:

$$C_2 = \frac{3}{2}\frac{\beta a^{1/2} V}{K_{\frac{1}{2}}(\beta a)}$$

$$C_4 = \frac{va^3}{2}\frac{K_{\frac{3}{2}}(\beta a)}{K_{\frac{1}{2}}(\beta a)}.$$

Substituting the values C_1, C_2, C_3, C_4 into formula (3.19)–(3.21), we obtain a solution to the problem posed for the current function and speeds in the form

$$\varphi = \frac{v}{2}\left(aR^3\right)^{\frac{1}{2}}\left[\frac{K_{\frac{1}{2}}(\beta R)}{K_{\frac{1}{2}}(\beta a)} - \frac{K_{\frac{3}{2}}(\beta R)}{K_{\frac{5}{2}}(\beta a)}\right]\sin^2(\theta) + \frac{va^3}{2R}\frac{K_{\frac{5}{2}}(\beta R)}{K_{\frac{1}{2}}(\beta a)}\sin^2(\theta) \tag{3.22}$$

$$v_R = \frac{v}{2}(a/R)^{\frac{1}{2}}\left[2\frac{K_{\frac{1}{2}}(\beta R)}{K_{\frac{1}{2}}(\beta a)} - \frac{K_{\frac{3}{2}}(\beta R)}{K_{\frac{5}{2}}(\beta a)}\right]2\cos(\theta) + \frac{va^3}{2R^5}\frac{K_{\frac{1}{2}}(\beta R)}{K_{\frac{1}{2}}(\beta a)}2\cos(\theta) \tag{3.23}$$

$$v_\theta = \frac{v}{2}(a/R)^{\frac{1}{2}}\left[2\frac{K_{\frac{1}{2}}(\beta R)}{K_{\frac{1}{2}}(\beta a)} - \frac{K_{\frac{5}{2}}(\beta R)}{K_{\frac{5}{2}}(\beta a)}\right]\sin(\theta) + \frac{va^3}{2R^3}\frac{K_{\frac{3}{2}}(\beta R)}{K_{\frac{5}{2}}(\beta R)}\sin(\theta). \tag{3.24}$$

Since, according to (3.19), the stream function and the Stokes operator of it are equal:

$$\varphi = \left\{\frac{C_1 R^{3/2}}{3\beta}\left[K_{5/2}(\beta R) - K_{1/2}(\beta R)\right] + C_1/R\right\}\sin^2(\theta)$$

$$D\varphi = C_2 R^{\frac{1}{2}} K_{\frac{3}{2}}(\beta R)\sin^2(\theta),$$

then from (3.4) it follows that

$$\begin{aligned}\partial P = \mu C_2\bigg[&\frac{1}{R^{1/2}}K_{3/2}(\beta R) - \frac{\beta}{3R^{1/2}}K_{5/2}(\beta R) + \frac{\beta}{3R^{1/2}}K_{1/2}(\beta R)\\ &-\frac{C_4}{C_2R^3}\beta^2\bigg]\cos\theta \mathrm{d}R - \bigg[\frac{1}{2R^{\frac{1}{2}}}K_{\frac{3}{2}}(\beta R) - R_{\frac{1}{2}}\beta\frac{\beta}{3R^{\frac{1}{2}}}K_{\frac{5}{2}}(\beta R)\\ &+\left(1-\frac{\beta R}{3}\right)\frac{R^{\frac{1}{2}}}{2}\beta K_{\frac{1}{2}}(\beta R) + \frac{C_4}{C_2R^2}\beta^2\bigg]\sin\theta \mathrm{d}\theta.\end{aligned} \tag{3.25}$$

Therefore, integrating (3.25), we obtain the expressions for pressure

$$P = P_0 + \frac{3}{2}\frac{\beta a^{\frac{1}{2}} v}{K_{\frac{1}{2}}(\beta a)}\left\{\frac{1}{2R}K_{\frac{3}{2}}(\beta R) + \left(1 - \frac{\beta R}{3}\right)\frac{R^{\frac{1}{2}}}{2}\beta K_{\frac{3}{2}}(\beta R)\right.$$

$$-\frac{a^{\frac{5}{2}}}{3R^2}\beta K_{\frac{3}{2}}(\beta R)\Bigg\}\cos\theta. \tag{3.26}$$

P_z of the liquid acting on the ball in the case under consideration is represented in the form

$$P_z \iint_{\Omega}\left(-P\cos\theta+\mu\frac{\partial w_z}{\partial R}\right)d\Omega \tag{3.27}$$

where w_z—component of the velocity vector parallel to the axis of symmetry; Ω—surface area.

This velocity component and its derivative with respect to radius R are determined as follows:

$$\begin{aligned} w_z &= v_R\cos\theta - v_\theta\sin\theta = \frac{ctg\theta}{R^2}+\frac{d\varphi}{\partial\theta}+\frac{1}{R}\frac{\partial\varphi}{\partial R}\\ \frac{\partial w_z}{\partial R} &= -\frac{2v_R\cos\theta}{R}+\frac{1}{R}v_\theta\sin\theta-\frac{ctg\theta}{R}\frac{\partial}{\partial\theta}(v_\theta\sin\theta)+\frac{1}{R}\frac{\partial^2\varphi}{\partial R^2}. \end{aligned} \tag{3.28}$$

Differentiating equation(3.19) we obtain

$$\begin{aligned}\left(\frac{\partial^2\varphi}{\partial R^2}\right)\downarrow_a &= \Bigg\{a^{1/2}C_2K_{3/2}(\beta a)-\frac{2}{3}\frac{C_2}{\beta a^{1/2}}K_{5/2}(\beta a)\\ &\quad+\frac{2}{3}\frac{C_2}{\beta a^{1/2}}K_{1/2}(\beta a)\frac{2C_4}{a^3}\Bigg\}\sin^2\theta\end{aligned} \tag{3.29}$$

Substituting the value of C_2, C_4 will have

$$\begin{aligned}\left(\frac{\partial^2\varphi}{\partial R^2}\right)\downarrow_a &= \frac{2}{3}\frac{\beta v a}{K_{1/2}(\beta a)}\Bigg\{K_{3/2}(\beta a)-\frac{4}{3}\frac{1}{\beta a^{1/2}}K_{5/2}(\beta a)\\ &\quad+\frac{2}{3}\frac{1}{\beta a^{1/2}}K_{1/2}(\beta a)\Bigg\}\sin^2\theta\end{aligned}$$

Using boundary conditions (3.5), we obtain

$$\left(\frac{\partial w_z}{\partial R}\right)=\frac{v\sin^2\theta}{a}+\frac{1}{a}\left(\frac{\partial^2\varphi}{\partial R^2}\right). \tag{3.30}$$

Substituting $\left(\frac{\partial^2\varphi}{\partial R^2}\right)$ into (3.30), we have

$$\left(\frac{\partial w_z}{\partial R}\right)=-\frac{v\sin\theta}{a}\left\{2+\frac{2}{3}\beta a\frac{K_{3/2}(\beta a)}{K_{1/2}(\beta a)}-2\frac{K_{5/2}(\beta a)}{K_{1/2}(\beta a)}\right\} \tag{3.31}$$

Thus, the integrand in (3.27) using recurrence relations for Bessel functions is equal to

$$\left(-P\cos\theta+\mu\frac{\partial w_z}{\partial R}\right)=-P_a\cos\theta-\frac{v}{a}\mu\left\{2+\frac{3}{2}\beta a\frac{K_{3/2}(\beta a)}{K_{1/2}(\beta a)}-2\frac{K_{5/2}(\beta a)}{K_{1/2}(\beta a)}\right\} \tag{3.32}$$

Calculating the integral and taking into account that

$$\iint\cos\theta\,\mathrm{d}s=0;\quad\iint\mathrm{d}s=4\pi a^2 \tag{3.33}$$

we'll find

$$P_z=-4va\mu\left\{2+\frac{3}{2}\beta a\frac{K_{\frac{3}{2}}(\beta a)}{K_{\frac{1}{2}}(\beta a)}-2\frac{K_{\frac{5}{2}}(\beta a)}{K_{\frac{1}{2}}(\beta a)}\right\}. \tag{3.34}$$

The last expression shows that, similar to the Navier–Stokes model, the resistance to the movement of the ball is proportional to the viscosity coefficient and the speed of movement.

In particular, when L_s has a small order, formula (3.34) approaches the well-known Stokes formula. Using the expressions of special functions [145] and taking $a=r$ (r—the radius of the ball under consideration), we obtain

$$\begin{aligned}
K_{3/2}(\beta r)&=\frac{1}{\beta r}K_{1/2}(\beta r)=K_{1/2}(\beta r)\left(\frac{1}{\beta r}+1\right)\\
K_{5/2}(\beta r)&=\frac{3}{\beta r}K_{3/2}(\beta r)+K_{1/2}(\beta r)\left(\frac{1}{\beta r}+1\right)\\
&=\frac{3}{\beta r}\left[K_{1/2}(\beta r)\left(\frac{1}{\beta r}+1\right)+K_{1/2}(\beta r)\right]+K_{1/2}(\beta r)\\
&=K_{1/2}(\beta r)\left[\frac{3}{\beta^2r^2}-\frac{3}{\beta r}-1\right].
\end{aligned}$$

From Eq. (3.34), the resistance force of a fluid or the force of influence of a fluid on a ball moving in it can be written in the form

$$\begin{aligned}
P_z&=-4\pi\mu rv\left\{2+\frac{3}{2}\beta r\frac{K_{\frac{1}{2}}(\beta r)}{K_{\frac{1}{2}}(\beta r)}\left(\frac{1}{\beta r}+1\right)-2\frac{K_{\frac{1}{2}}(\beta r)}{K_{\frac{1}{2}}(\beta r)}\left(\frac{1}{\beta r}+1\right)\right\}\\
&=-4\pi\mu rv\left[2+\frac{3}{2}+\frac{3}{2}\beta r-\frac{6}{\beta^2r^2}-\frac{6}{\beta r}-2\right].
\end{aligned} \tag{3.35}$$

Or denoting $P_z = -R_c$, we finally obtain the following new entry for the expression of the resistance force:

$$R_c = 6\pi\mu r v\left[1 + \beta r - \frac{4}{\beta^2 r^2} - \frac{4}{\beta r}\right] \tag{3.36}$$

The obtained dependence allows us to take into account the influence of the mode of movement of suspended particles in a fluid flow on the process of sediment transport. Thus, this dependence has a fundamental advantage over the existing formulas of Stokes, Oseen, Goldstein, and other researchers.

Thus, based on the model (Chap. 2) of fluid movement, formula (3.6) is obtained, which allows us to determine the resistance force R_c the movement of a solid particle in a liquid and in laminar and turbulent modes of motion.

3.1.1 Modes of Movement of Suspended Particles in a Fluid Flow

As noted above, determining the characteristics of suspended sediment particles is of great practical and theoretical importance, especially in describing the nature of the flow itself transporting particles suspended in it. Determination of the coefficient of any resistance of sediment particles is required to correctly account for the force of interaction between the carrier fluid and suspended particles, as well as when determining individual characteristics of suspended particles.

For these purposes, numerous theoretical and experimental studies have been carried out [4, 31, 37, 39, 43, 47, 51, 68,1 38].

When determining the coefficient of any resistance of solid particles, their size plays an important role. Depending on the size, the following modes of movement of solid particles in a liquid are distinguished:

(1) laminar ($\mathrm{Re}_d < = 1$), in this mode the value of the coefficient of any resistance of solid particles reduced to the shape of a ball can be expressed using the Stokes formula [4] in the form

$$C_0 = \frac{24}{\mathrm{Re}_d} \tag{3.37}$$

(2) turbulent ($d > 0.0015$ m and $\mathrm{Re}_d > 1000$ or area of square resistance). Here the value of C_0 is determined by the well-known Newton-Rittenger formula:

$$\overline{R_c} = C_0 \rho F \frac{u^2}{2},$$

where ρ—density of the liquid medium;

F—cross-sectional area of particles;

u_0—relative speed.

In this area C_0 is assumed to be constantly equal $C_0 = 0.46$ [4, 37, 43, 47, 142].

(3) intermediate $0.000015 < d < 0.0015$ m and $1 < \mathrm{Re}_d < 1000$.

For this region, the search for a reliable equation for C_0 has great practical significance.

It should be noted that the movement of river sediments often takes place within the laminar and intermediate regime, since in most cases river sediments consist of those fractions whose diameter varies within $0.000015 < d < 0.0008$ m, hence the special interest in determining C_0 for the specified modes.

When studying the movement of a solid particle in a flow, it is also necessary to take into account the interaction of solid particles with surrounding particles and the difference in the shape of the particle from spherical [37, 43, 47, 138, 165]. The material presented in these works allows us to conclude that in the case of low concentrations, the mutual influence of particles can be neglected, and in the case of large concentrations, they can be taken into account, corrected for constraint.

Detailed measurements of lift and drag acting on a cylinder in a liquid at various distances from a solid surface were carried out by M. A. Dementiev [52].

Based on experiments, M. A. Dementyev came to the conclusion that the lifting force decreases as the cylinder moves away from the solid surface and disappears at a distance almost equal to the diameter of the particle in question. Similar conclusions are given in works [139, 185, 189]. Here the conclusions about the constancy of the value of C_0 are confirmed in the region of quadratic resistance and the numerical value of C_0 is determined = 0.46.

V. K. Debolsky [49], based on processing numerous experimental data to determine the drag coefficient for various modes, proposes the following formulas:

$$\begin{array}{ll} \mathrm{Re}_d < 1, & C_0 = \frac{24}{\mathrm{Re}_d}\left(1 + \frac{1}{4}\mathrm{Re}_d - \frac{7}{400}\mathrm{Re}_d^2\right) \\ 1.0 < \mathrm{Re}_d < 800, & C_0 = \frac{24}{\mathrm{Re}_d}\left(1 + \frac{1}{6}\mathrm{Re}_d\right) \\ 800 < \mathrm{Re}_d < 10^6, & C_0 = 0.5 \end{array} \tag{3.38}$$

The work [36] considers the value of the coefficient of any resistance for solid particles during constrained settling and provides a formula for determining C_0 under the specified conditions.

For laminar motion of solid particles in a liquid, the formula for determining C_0 for any isometric particle shape is considered in the form

$$C_0 = \frac{24.47}{\mathrm{Re}_d \lg \frac{15.3g}{\theta}} \tag{3.39}$$

where θ is the coefficient of non-sphericity of particles, equal to the ratio of the surface of the particle to the surface of a ball of equal size.

Oseen, partially using the inertial terms of the equation of motion of a viscous fluid, to determine C_0 offers a formula in the form

$$C_0 = \frac{24}{\mathrm{Re}_d}\left(1 + \frac{1}{16\mathrm{Re}_d}\right). \tag{3.40}$$

Thus, the above suggests that searching for the value of C_0 depending on the Reynolds number (Re_d) is the most appropriate.

Let's try to determine the coefficient of any resistance of sediment particles based on formula (3.36)

$$R_c = -6\pi\mu r v\left[1 + \beta r - \frac{4}{\beta^2 r^2} - \frac{4}{\beta r}\right] \tag{3.41}$$

where $\beta = (L_s/\mu)^{1/2}$.

From (3.42), the drag coefficient of suspended particles reduced to the shape of a ball, and the dependence on the Reynolds number can be expressed as follows:

$$C_0 = \frac{24}{\mathrm{Re}_d}\left[1 + \beta r - \frac{4}{\beta^2 r^2} - \frac{4}{\beta r}\right]. \tag{3.42}$$

Now we find expressions βr. If we express it as a function Re_d, then, limiting ourselves to two terms of this series, we have

$$\beta r = (A + B \cdot Re_d). \tag{3.43}$$

The parameters of Eq. (3.44) A, B are determined by simultaneous solution of Eqs. (3.43), (3.44) under the following boundary conditions:

$$\mathrm{Re}_d = 1,0; \qquad C_0 = 28.0$$
$$\mathrm{Re}_d \geq 1000;\ C_0 = \mathrm{const} = 0.46$$

Consider the case $\mathrm{Re}_d \rightarrow$ then we write (3.44) in the form

$$\beta r = \beta \mathrm{Re}_d. \tag{3.44}$$

Substituting (3.45) into (3.43), we have

$$0,46 = \frac{24}{\mathrm{Re}_d} = \left[1 + B\mathrm{Re}_d - \frac{4}{B\mathrm{Re}_d} - \frac{4}{B^2\mathrm{Re}_d^2}\right],$$

discarding terms with small values, we obtain $B = 0.018$. When $\mathrm{Re}_d = 1$; $C_0 = 28$, then from (3.43) $\beta r = 2.43$.

From here, finally for βr, we get the following formula

$$\beta r = (2.45 + 0.018\mathrm{Re}_e) \tag{3.45}$$

It should be noted that this formula also allows one to determine expressions for L_s—a coefficient that takes into account turbulent mixing of suspended particles in a liquid flow in the form:

$$\beta = (L_s/\mu)^{1/2}$$
$$L_S = 4\mu\frac{(2.45 + 0.018Re_d)^2}{d^2}. \tag{3.46}$$

The resulting formula (3.43) was compared with the measured data given in [4, 33, 47, 148]

Figure 3.2 shows the dependence curves $C_0 = f(\text{Re}_d)$ according to the Stokes, Oseen formulas, and (3.43), as well as according to the empirical dependence of V.K. Debolsky.

The calculation results are summarized in Table 3.1.

As can be seen from the graph, comparisons of measured and calculated data give satisfactory results.

Thus, a generalized dependence is proposed for determining the drag coefficient of suspended sediment particles, which allows one to determine the value of C_0 in the case when $\text{Re}_d \gg 1$.

On the other hand, according to the proposed dependence, the values of the drag coefficient for all driving modes are determined by a single formula, which greatly simplifies the use of the formula for practical calculations.

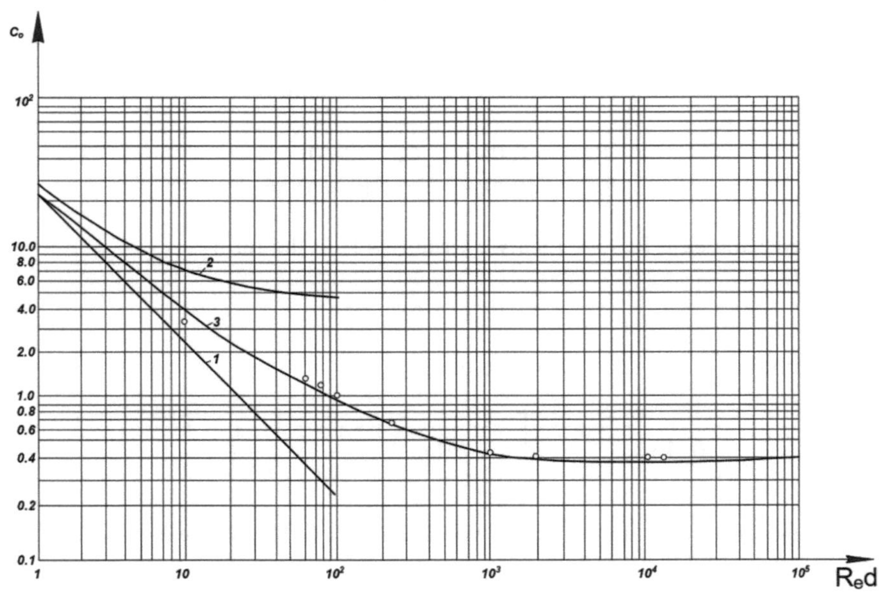

Fig. 3.2 Dependence of the drag coefficient of sediment particles on the Reynolds number according to the formulas of Stokes (1), Oseen (2), Arifjanov (3)

Table 3.1 Towards the calculation of the drag coefficient of sediment particles

No. pp	Re_d	βr	$\frac{4}{\beta r}$	$\frac{4}{\beta^2 r^2}$	According to formula (3.43)	According to the formula of V. K. Debolsky
1	0, 1	2.452	1.63	0.665	257.0	234
2	1.0	2,468	1.62	0.656	24.60	19.8
3	10	2.63	1.52	0.57	3.78	4.2
4	50	3.35	1.19	0.356	1.35	1.5
5	80	3.89	1.03	0.264	1.08	1.20
6	100	4.25	0.94	0.22	0.98	1.07
7	200	6.05	0.66	0.109	0.75	0.81
8	500	11.45	0.349	0.0305	0.57	0.6
9	800	16.85	0.237	0.014	0.51	0.45
10	1000	20.45	0.195	0.0095	0.509	0.5
Eleven	3×10^3	56.45	0.0708	0.00125	0.46	0.5
12	3×10^4	182.45	0.022	0.00012	0.44	0.5
13	5×10^4	902.45	0.0044	0.000005	0.43	0.5
14	1×10^5	1802.45	0.0022	0.000001	0.43	0.5

3.2 Hydraulic Sediment Size

Hydraulic size is one of the main parameters of suspended sediment particles, characterizing the interaction of a solid particle with a carrier fluid.

Many researchers have studied hydraulic size [7, 31, 39, 49, 50, 67, 68, 75, 138, 163, 185, 189, 195].

The first theoretical developments in this direction belong to D. Stokes (1850). Stokes studied the fall of small fractions of particles in stagnant water with a very low relative speed. According to Stokes, the resistance force when a spherical particle falls in a liquid is determined as follows:

$$R_c = 3\pi \mu w d \tag{3.47}$$

where d—diameter of a spherical particle;

w—hydraulic particle size.

With a uniform fall, gravity F_t balanced by the resistance force R_C.

$$F_m - R_c = 0$$

Let us express the force of gravity of a spherical particle in a liquid as follows:

$$F_m = \pi d^3 g(\rho_m - \rho)/6$$

where p, p_m—the density of the liquid and the particle, respectively.

Substituting the values of gravity, we obtain a formula for determining the hydraulic size, with a uniform fall of small fractions of solid particles:

$$3\pi\mu wd = \pi d^3 g(\rho_m - \rho)/6 \tag{3.48}$$

$$w = \frac{k_1 g d^2(\rho_m - \rho)}{4\nu\rho}. \tag{3.49}$$

where k_1—numerical dimensionless coefficient depending on the shape of the particles; for spherical particles $k_1 = 0.22$ [73, 131].

Further experimental tests of formula (3.49) showed that this formula is suitable for $\mathrm{Re}_e \leq 1$, i.e. in laminar deposition mode.

The most detailed experimental studies of hydraulic fineness were carried out by A. P. Z. Egzhda, B. V. Arkhangelsky, V. N. Goncharov, V. V. Romanovsky, V. K. Debolsky, and others. Research by the above-mentioned authors showed that the value of hydraulic fineness depends on the deposition mode, particle size and shape, liquid temperature, etc.

An important point occurs when a solid particle moves in a turbulent regime.

V. N. Goncharov suggests taking this factor into account using the turbulence coefficient. This parameter varies from 205 to 1 depending on the diameter of the solid particle.

Thus, depending on the flow characteristics, the hydraulic fineness changes on the order of a hundred times.

The peculiarity of the experimental studies conducted by V. V. Romanovsky [131], is that he studied the falling speed of a solid particle as a function of the mass, size, and shape of the particle. As a result, he published a new scale of hydraulic particle size. In the turbulent mode of sedimentation of a solid particle in a liquid flow, V. V. Romanovsky proposes the following relationship for determining the hydraulic size:

$$w = k_1\left(\frac{gd(\rho_m - \rho)}{\rho}\right)^{\frac{1}{2}}. \tag{3.50}$$

A feature of this formula is that it takes into account the coefficient k_1, which was previously established. In the absence of data on particle shape, the following formula is proposed to determine the hydraulic size:

$$w = 1.22\left(\frac{gd(\rho_m - \rho)}{\rho}\right)^{\frac{1}{2}}.$$

Theoretical and experimental studies conducted by V. K. Debolsky made it possible to express the formula for hydraulic size in the form:

$$w = \left(4gd\rho_0/\left(3C_0\overline{\theta}\right)\right)^2 \tag{3.51}$$

where $\rho_0 = (\rho_m - \rho)/\rho$;

$\overline{\theta}$—coefficient taking into account the shape of suspended particles.

Unlike previous formulas, formula (3.52) takes into account the deposition mode and shape of solid particles.

We are considering the definition of the expression for hydraulic fineness using from formula (3.46). The equations for changing the kinetic energy of a particle in the vertical plane are written as:

$$mdw^2 = 2(F_m - R_c)\mathrm{d}y$$

where $R_c = C_0 Fpw^2/2$—the resistance force in the turbulent regime of falling solid particles;

m $= \rho_m V_m$—mass of solid particle.

$$\rho_m V_m dw^2 = 2\left[V_m g(\rho_m - \rho) - C_0 F \rho w^2/2\right]\mathrm{d}y$$

Separating the variables and integrating we find:

$$w = \left(2V_m g(\rho_m - \rho)\left(1 - \exp\left(-\frac{CF\rho y}{(\rho_m V)}\right)\right)\right)/CR\rho)^{1/2}. \tag{3.52}$$

At $y \to \infty$ Eq. (3.53) will take the form

$$w = \left(\frac{2gd(\rho_m - \rho)}{C_0\rho}\right)^{\frac{1}{2}}. \tag{3.53}$$

Now, expressing C_0 through the Reynolds number, we obtain the following dependence for the hydraulic size:

$$w = kgd^2(\rho_m - \rho)/(4\nu\rho)^2 \tag{3.54}$$

$$k = 0.22/\left[1 + \beta r - \frac{4}{\beta^2 r^2} - \frac{4}{\beta r}\right] \tag{3.55}$$

where k is a coefficient depending on the shape of the particle and the mode of sedimentation of the solid particle in the liquid flow.

Note that, on the basis of numerous experiments, V. V. Romanovsky [132] established that in laminar sedimentation mode, the hydraulic particle size does not depend on the particle shape and the sizes of the measured sediment particles were within the range of 0.004–0.05 mm. A comparison of the measured data of V. V. Romanovsky with the results of calculations using formula (3.55) in the indicated mode gives satisfactory results.

The peculiarity of the proposed formula is that it makes it possible to identify the boundaries of the mode of movement of a solid particle in a fluid flow, and also generalizes existing formulas that determine the hydraulic size of sediment.

Table 3.2 Modes of motion of a solid particle in a liquid flow

No. /p	Re_d	βr	$\frac{4}{\beta r}$	$\frac{4}{\beta^2 r^2}$	k
1	0.1	2.452	1.63	0.665	0.2
2	1.0	2,468	1.62	0.656	0.22
3	10	2.63	1.52	0.57	0.16
4	50	3.35	1.19	0.356	0.08
5	100	4.25	0.94	0.22	0.054
6	200	6.05	0.66	0.109	0.031
7	500	11.45	3.349	0.0305	0.016
8	800	16.85	0.237	0.014	0.012
9	1000	20.45	0.195	0.0095	0.011

In the laminar mode of particle deposition in a fluid flow, which corresponds to $\mathrm{Re}_d \leq 1$, we obtain the known value of the coefficient $k = 0.22$. The table below shows the change in coefficient k depending on the mode of liquid flow around a solid particle, calculated using formula (3.56) (Table 3.2).

This table allows us to draw a conclusion about the boundaries of the regime of sedimentation of a solid particle in a liquid flow and theoretically substantiate the dependence of the value of the hydraulic particle size on the regime of liquid flow around it. When determining the values of hydraulic fineness, it becomes necessary to take into account the shapes of grains of transporting sediment. The parameter characterizing the shape of particles is the coefficient of non-sphericity of particles $\overline{\theta}$, equal to the ratio of the surface of the particle to the surface of a ball of equal volume. Taking into account the above, we write expression (3.55) in the form $w = kgd^2\rho_0/(4\nu\rho\log(15.4/\theta))$, where $\rho_0 = (\rho_m - \rho)/\rho$. An analysis of this formula is shown in Fig. 3.3, which shows the dependence of the relative hydraulic size on their shape and the Reynolds number Re_d. $w = \varphi\left(Re_d, \overline{\theta}\right)(\rho_0 gd)^2$. Experimental data are plotted on the same graph [138]. The comparison results show satisfactory agreement between the measured and calculated data (Table 3.3).

In order to analyze the possibility of using the proposed formula (3.55) for a large range of changes in the size and density of suspended particles, the calculated data were compared with the experimental data of V. K. Debolsky [47]. The comparison results are given in Table 3.4. Analysis of the calculated and measured results confirms the possibility of using formula (3.55) in engineering calculations in a wide range of changes in the size and density of suspended particles.

3.2.1 Hydraulic Fineness of Suspended Sediment Particles in Cramped Conditions

Now let's consider the possibilities of using formula (3.55) in cramped conditions. Researchers have established that the dependences for hydraulic size during the

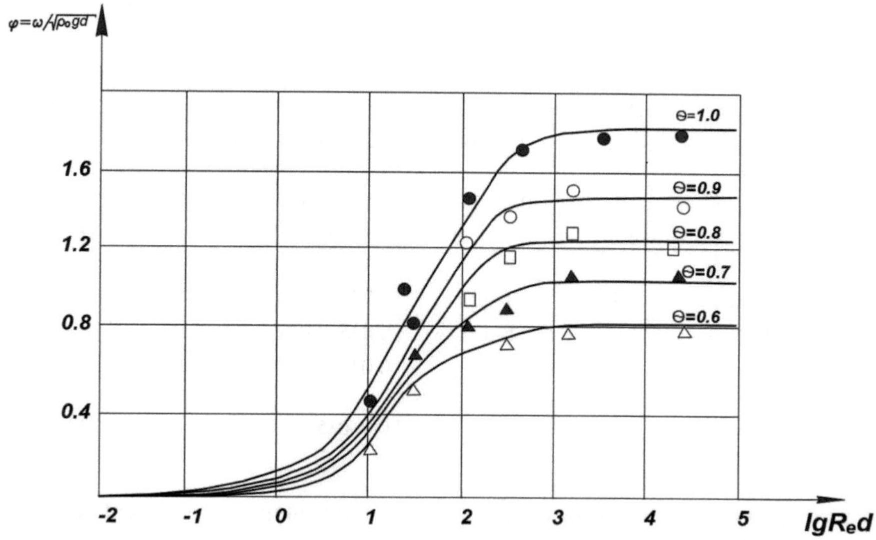

Fig. 3.3 Dependence of the relative hydraulic size of particles on their shape and Reynolds number

Table 3.3 .

Re_d	$\overline{\theta}$				
	1.0	0.9	0.8	0.7	0.6
10	0.4	0.32	0.31	0.30	0.29
50	1.0	0.81	0.77	0.75	0.71
100	1.25	1.1	0.97	0.8	0.75
200	1.6	1.31	1.2	1.1	0.79
500	1.75	1.43	1.3	0.9	0.795
1000	1.8	1.47	1.39	1.45	0.8

movement of single particles in a boundless environment and during sedimentation in suspension, i.e. in a flow of aggregate particles (in cramped conditions) are different. There are many ways to take into account the deposition of solid particles in cramped conditions [31, 47, 68, 74, 86, 94, 143, 191, 193, 195].

M. A. Velikanov proposes to define the expression for hydraulic size in cramped conditions w_0 in the following form:

$$w_0 = (1 - s)$$

where s—volumetric content of suspended particles.

However, further research in this direction showed that this formula requires additional clarification. V. K. Debolsky proposes to determine the hydraulic size

Table 3.4 Results of comparison of calculated and measured values of hydraulic fineness according to laboratory data of V. K. Debolsky

No. p /p/	Material	Density $pg \cdot s\ \mathrm{m}^3$	Fraction size d, mm	K	Calculated w, cm/s	Measured w, cm/ s
1	Lyubertsy sand	2.65	0.1–0.6	0.25–0.01	1.01–5.83	1.0–6.0
2	Volsky sand	2.65	0.15–1.0	0.17–0.035	1.54–14.16	1.5–14.0
3	Fireclay powder	2.21	0.06–1.0	0.55–0.035	0.58–10.40	0.6–11.0
4	Bakelite	1.41	0.1–1.0	0.25–0.035	0.26–3.54	0.3–3.5
5	Polystyrene	1.05	2.5–1.0	0.045–0.011	3.21–4.95	3.5–5.0
6	Magnite sand	4.45	0.8	0.042	23.4	25.0
7	Pebbles	2.50	433	0.00035	238	230

depending on the concentration and mode of movement of suspended sediment particles in the form:

$$w_0 = (1 - s)^{\varphi_2}$$

where is φ_2—the coefficient, the value of which is taken depending on the Reynolds number [128].

Taking into account the above expression, to determine the hydraulic size in cramped conditions, we can write

$$w_0 = kgd^2(\rho_m - \rho)(1 - s)^{\varphi_2}/\nu\rho.$$

Figure 3.4 shows the results of an analysis of the calculated and measured values of the relative particle size depending on their concentration according to data from different authors (B. Kay, R. Boardman, P. Cheng, T. Shahman, J. M. Aknoun, P. Lin, V. K. Debolsky).

The hydraulic fineness value is also affected by the presence of silt particles. Due to the saturation of the flow with small clay particles, the density and viscosity of the mixture changes differently by different researchers [4, 49, 78, 108]. The most convincing evidence in this regard was expressed by the authors of [102], where it is proposed to divide the sediment transported by the flow into two parts: the first, suspended sediment, the diameter of which is more than 0.015 mm; second sediments, the diameter of which is less than 0.015 mm. The presence of the former is determined by formulas that determine the transporting capabilities of the flow.

The presence of the latter is taken into account through the density and viscosity of the carrier flow. The density of a turbid medium is determined by the equality

$$\rho_c = \rho(1 - s) + \rho_m s.$$

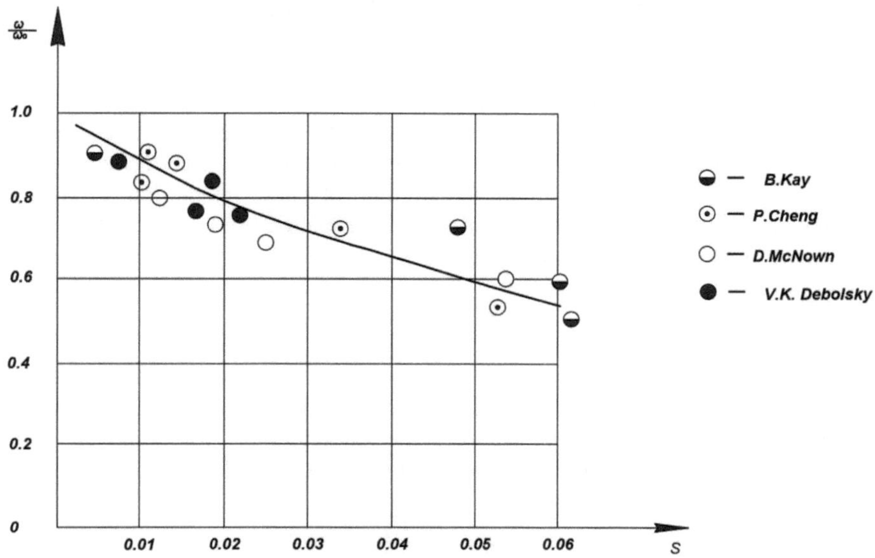

Fig. 3.4 Changes in relative hydraulic particle size as a function of concentration

Determining the influence of clay particles on the viscosity of the carrier flow has been studied by many researchers. The most generalized expression for determining the viscosity of a suspension was proposed by A. Einstein [189]

$$\mu_c = \mu[1 + 2.5(1 - s)]$$

Kh. A. Ismagilov [74], studying mudflows to determine the coefficient of kinematic viscosity, proposes the following dependence

$$v_c = v(1 + 1.1s)(d_1/(2d_2))^{0.65}$$

where d_1—upper limit of clay particles;
d_2—lower limit of clay particles.

Following the above, using relations (3.55), we write the generalized dependence for determining the hydraulic size in the form

$$w = kgd^2(\rho_m - \rho_c)(1 - s)^{\varphi_2}/(v_c\rho_c).$$

Thus, a calculation formula has been proposed that generalizes a number of factors characterizing the movement of a solid particle in a liquid flow, which makes it possible to determine the hydraulic particle size under various sedimentation modes and isometric forms of solid particles, as well as in cramped conditions, taking into account the presence of clay particles that have of particular importance in the study of suspended flow in channels.

3.3 Hydrodynamic Interaction of Suspended Particles with Liquid

As noted above, one of the main elements of a suspension-carrying flow when considering them separately (carrying liquid and solid particle) is taking into account the force of interaction between the phases. Researchers have proposed various expressions for determining the strength of phase interaction. In general, as noted above, the strength of interaction depends on the rates of each phase, their derivatives, concentration, particle sizes, etc.

$$\Pi = f\left(U, U_S, \frac{dy_i}{dx_i}, s, d\right)$$

F. I. Frankl [156–158] when separately considering the suspension flow, i.e. when drawing up equations of averaged motion separately for liquid and solid particles, the interaction forces are proposed to be expressed as follows:

$$\Pi_i = s' \frac{\partial P_{ik}}{\partial x_k};$$

where P_{ik}—tensor of stresses caused by molecular interactions.

However, expressing the interaction force in this form does not take into account the range of parameters characterizing the interaction between the carrier fluid and the solid particle, and such an entry does not make it possible to close the proposed equations.

Further, F. I. Frankl proposes a method for closing the equations of motion as follows: the fluid flow field is considered known and independent of the presence of suspended sediment particles. Then the expressions for the interaction force are written as:

$$\Pi_i = -\frac{9}{2} s' \frac{\mu}{r^2} (u_{si} - u_i - U_{si})$$

where u_i—fluid velocity components;

$$u_{si} = -\varepsilon \frac{\partial(\ln s)}{\partial x_1}$$

ε_1—turbulent kinematic viscosity.

The difficulty of using expressions (3.58) for practical calculations lies in the fact that the formula contains a number of unknown parameters (u_{si}, ε_1).

The definition of which requires additional research.

Studies by a number of authors in this direction [95, 99, 120, 153] have shown the possibility of expressing the interaction force P as a proportional difference in the speeds of each phase, i.e. in vector notation

$$\Pi = K(\overline{u} - \overline{u_s}), \tag{3.56}$$

where $\overline{u}$, $\overline{u_s}$—velocities of the carrier fluid and solid particles;
K—*phase interaction* coefficient.

The accepted hypothesis (3.59) for the expression P is so far the most acceptable, as evidenced by the solution of practical problems [93,120,148,155,163, etc.].

When small particles corresponding to the Stokes region move in a liquid, the interaction force will be written as:

$$\Pi = 18\mu s(u - u_s)/d^2.$$

With further changes in the size of suspended particles, i.e. in the transition and turbulent region of resistance, expressions for the interaction force are determined depending on the value of the phase interaction coefficient.

3.3.1 About the Phase Interaction Coefficient

To determine the value of the interaction coefficient K, theoretical formulas and experimental data are used. In the case when the flow consists of a liquid and a solid particle (water + rosin), experiments to determine K were carried out by the authors [95] in the following ways:

(a) by measuring the friction acting on a plate moving along the surface of a medium;
(b) by measuring the volumetric flow rates of each phase during the flow of two-phase media in a horizontal cylindrical pipe.

These experiments were carried out at a constant concentration and low Reynolds numbers, as a result of which the resulting value of the phase interaction coefficient was a constant value.

Under conditions of channel flows, the task becomes much more complicated, since the concentration and size of solid particles are distributed unevenly across the depth, increasing from the surface to the bottom. Therefore, the expression for determining K under conditions of channel flows must take into account the uneven distribution of concentration (turbidity) and size of solid particles along the depth of the suspended flow.

$$R_c = C_0 \rho \pi d^2 (\overline{u} - \overline{u_s})^2/8 \tag{3.57}$$

where C_0—particle drag coefficient;
d—particle diameter;
ρ—density of the carrier fluid.

If the number of particles in the elementary volume W is equal to n, then the total resistance force

$$\Pi = \frac{nR_c}{W}. \tag{3.58}$$

Let us denote by W the volume of particles in the elementary volume, then we write (3.61) in the form:

$$\Pi = \frac{n\overline{R}_c}{W_s}\frac{W_s}{W} = \frac{nR_e}{W_s}s \tag{3.59}$$

knowing that $\frac{W_s}{W} = s$—the concentration of suspended particles.

Because

$$W_s = \pi d^3 n/6$$

then the interaction force is equal

$$\Pi = \frac{3C_0 s\rho(\overline{u} - \overline{u}_s)^2}{4d}. \tag{3.60}$$

From here, using the phase interaction coefficient, we determine it as follows:

$$K = \frac{3C_0\rho(\overline{u} - \overline{u}_s)^2}{4d}. \tag{3.61}$$

Using (3.37), (3.40), (3.41), and (3.43), K can be expressed as follows:
in laminar mode

$$K = 18\mu s/d^2.$$

According to Oseen's formula (with partial allowance for the nonlinearity of the nature of motion)

$$K = 18\mu s(1 + 1/(16Re_d))/d;$$

according to formula (3.40) (with constrained movement of particles in suspension)

$$K = 18{,}095 \cdot \mu s/d^2;$$

according to formula (3.43), a generalized dependence for determining the phase interaction coefficient

$$K = 18\mu s\left(1 + \frac{\beta d}{2} - 8/(\beta d) - 16/(\beta^2 d^2)\right)/d^2.$$

Consequently, the equation for the force of interaction between a carrier fluid and a solid particle (3.59), for various modes of motion of the latter, can be written in the form

$$\Pi = 18\mu s\left(1 + \frac{\beta d}{2} - 8/(\beta d) - 16/(\beta^2 d^2)\right)(u - u_s)/d^2.$$

Thus, a qualitatively new dependence is proposed for determining the strength of interaction between phases, which allows one to evaluate the interactions of the carrier liquid and solid particles even at high Reynolds numbers, taking into account the size of suspended particles and the uneven distribution of the concentration of suspended particles along the depth of the flow.

4 Kinematic Characteristics of Suspended Flow

4.1 Velocity Field of a Turbulent Suspension-Carrying Flow

A more accurate description of the pattern of distribution of average velocities over the depth of the flow is the most convincing confirmation of the correctness of the accepted theoretical premises.

Velocity profiles in open water flows remain poorly studied, even with single-phase flow. Universal shapes of velocity profiles can be found using various hypotheses.

In the process of studying the movement of fluid in various years, many formulas of an empirical or semi-empirical nature were proposed for calculating the velocity distribution over the flow cross section. In single-phase flows, determining the distribution of average velocity over the depth of the flow is the basis for estimating the resistance to flow movement in the channel. And some progress has been made in the study of this phenomenon.

Existing formulas that determine the distribution of velocity over the depth of the flow can be classified as one of the following types: parabolic, logarithmic, exponential (power), and elliptical.

In the most simplified form, these formulas are expressed as:

$$u = u_{\max} - \frac{m}{C} u_c (1 - \eta)^2 \quad \text{parabolic;} \tag{4.1}$$

$$u = u_{\max} + \frac{V_z}{æ} \ln \eta \quad \text{logarithmic;} \tag{4.2}$$

$$u = u_{\max} * \eta^x \quad \text{power;} \tag{4.3}$$

$$u = u_{\max} \sqrt{1 - P\eta^2} \quad \text{elliptical;} \tag{4.4}$$

© The Author(s), under exclusive license to Springer Nature Singapore Pte Ltd. 2026

A. M. Arifjanov et al., *Dynamics of Suspended Flow in Open Channels*,
https://doi.org/10.1007/978-981-96-8781-7_4

where

u_{max}—surface velocity;
u—local speed at a distance from the bottom;
u_c—average vertical speed;
V_z—dynamic speed;
C—Chezy coefficient;
η—relative depth,

m, æ, x, P—respectively, the parameters of the parabolic, logarithmic, power, and elliptic velocity profiles.

These parameters are mainly determined based on experiments and vary over large ranges.

A detailed analysis of the theoretical premises of the velocity distribution formulas in the case of a single-phase flow is considered in [5, 19, 23, 32, 42, 50, 79, 180].

It should be noted that in suspension-carrying flows, the nature of the velocity distribution along the depth of the suspension-carrying flow becomes significantly more complicated, since it depends not only on the present liquid and solids, etc. The proposed formulas for describing the nature of the distribution of average velocities in suspension-carrying flows are based on the well-known Prandatl-Karman hypothesis for homogeneous liquids [9, 20, 32, 34, 42, 49, 69, 77, 78, 105, 111, 115, 150, 164, 188, 195, 199, 200] and assessment of the influence of suspension on the velocity distribution carried out through some parameters.

However, the proposed parameters, which take into account the influence of a solid particle on the velocity field, are in complex dependence on the characteristics of suspended particles and were obtained experimentally for specific conditions, which naturally narrows the scope of their application.

Modern advances in science make it possible to solve the above problem taking into account many factors characterizing the nature of the suspended flow. In particular, taking into account the interaction of suspended particles and the carrier fluid, the uneven distribution of the concentration of suspended particles and their fractional composition.

Certain works in this direction are given in works [9, 143, 144], which show the results of a more complete experimental study of the nature of suspended flows in pipelines. However, it should be taken into account that the dependencies given in these works were obtained for the case of the movement of suspension-carrying flows in pressure systems and in order to use them for open flows, it becomes necessary to take into account the flow features. And all this points to the search for new dependencies to determine the distribution of the average velocity over the depth of the suspended flow.

4.1.1 Method for Calculating the Distribution of Averaged Velocity over the Depth of a Suspended Flow

To develop a method for calculating the distribution of average velocity over the depth of a suspension-carrying flow, we proceed from the system of differential Eqs. (2.14). In the case of steady $\left(\frac{du}{dt}=0\right)$ and uniform flow movement, the system of differential equations (2.14) will be written in the form:

$$\left.\begin{array}{l}\text{for carrier liquid:}\\ \frac{\partial}{\partial y}\left[(1-s)\mu\frac{\partial u}{\partial y}\right]+K(u_s-u)-(1-s)Lu=-(1-s)\rho gi\\ \text{For suspended particles:}\\ \frac{\partial}{\partial y}\left[s\mu\frac{\partial u_s}{\partial y}\right]+K(u-u_s)-sL_su_s=-s\rho_m gi.\end{array}\right\}\quad(4.5)$$

Now let's consider solving the system of differential equations (4.5) using the numerical finite difference method. Then, from the differential equation (4.5) we proceed in the usual way to the finite-difference equation.

We will consider solving the equation under the following boundary conditions:

$$u_1=0;\quad u_{s1}=0;$$

at $y=0\; j,k=1,2,3\ldots;$

$$u_j^{k+1}=u_{j-1}^{k+1};\quad u_{sj}^{k+1}=u_{sj-1}^{k+1};$$

$$\begin{aligned}\frac{\partial}{\partial y}\left[(1-s)\mu\frac{\partial u}{\partial y}\right]&=\frac{\mu}{\Delta y^2}\left[\left(1-s_{j+\frac{1}{2}}\right)(u_{j+1}-u_j)-\left(1-s_{j-\frac{1}{2}}\right)(u_j-u_{j-1})\right]\\ &=\frac{\mu(1-s_{j+0,5})}{\Delta y^2}u_{j+1}+\frac{\mu(1-s_{j-0,5})}{\Delta y^2}u_{j-1}-\frac{\mu(1-s_{j+0,5}-s_{j-0,5})}{\Delta y^2}u_j.\end{aligned}\quad(4.6)$$

Therefore, we write for the other terms of Eq. (4.5):

$$\left.\begin{array}{l}\text{for carrier liquid:}\\ \frac{\mu(1-s_{j+0,5})}{\Delta y^2}u_{j+1}+\frac{\mu(1-s_{j-0,5})}{\Delta y^2}u_{j-1}-\left[\frac{\mu(2-s_{j+0,5}-s_{j-0,5})}{\Delta y^2}\right.\\ \quad\left.+K+(1-s_j)L\right]u_j+Ku_{sj}+(1-s_j)\rho gi=0\end{array}\right\}\quad(4.7)$$

for suspended sediment particles:

$$\begin{aligned}&\frac{\mu s_{j+0,5}}{\Delta y^2}u_{sj+1}+\frac{\mu s_{j-0,5}}{\Delta y^2}u_{sj-1}-\left[\frac{\mu(s_{j+0,5}-s_{j-0,5})}{\Delta y^2}+K+s_jL_s\right]u_{sj}\\ &\quad+Ku_j+s_j\rho_T gi=0.\end{aligned}$$

Introducing the notation:

$$A = \frac{\mu\left(1 - s_{j+0,5}\right)}{\Delta y^2}; \qquad A_S = \frac{\mu s_{j+0,5}}{\Delta y^2};$$

$$B = \frac{\mu\left(1 - s_{j-0,5}\right)}{\Delta y^2}; \qquad B_S = \frac{\mu s_{j-0,5}}{\Delta y^2};$$

$$c = \frac{\mu\left(2 - s_{j+0,5} - s_{j-0,5}\right)}{\Delta y^2} + K + \left(1 - s_j\right)L;$$

$$C_s = \frac{\mu\left(s_{j+0,5} - s_{j+0,5}\right)}{\Delta y^2} + K + s_j L_s;$$

$$D = \left(1 - s_j\right)\rho g i + K u_{sj}; \qquad D_s = s_j \rho_t g i + K u_j,$$

we write expression (4.7) in the form:

$$\left.\begin{aligned} A u_{j+1}^{k} + B u_{j-1}^{k+1} - C u_j^{k+1} + D = 0 \\ A_S u_{sj+1}^{k} + B_s u_{sj-1}^{k+1} - C_s u_{sj}^{k+1} + D_s = 0, \end{aligned}\right\} \tag{4.8}$$

hence the expression:

For carrier fluid speed:

$$u_j^{k+1} = \frac{A u_{j+1}^{k} + B u_{j-1}^{k+1} + D}{C}; \qquad 1 \le j \le j - 1 \tag{4.9}$$

for the speed of suspended particles:

$$u_j^{k+1} = \frac{A_S u_{sj+1}^{k} + B_s u_{sj-1}^{k+1} + D_s}{C_s}; \qquad 1 \le j \le j - 1. \tag{4.10}$$

Calculation operations were carried out on a computer, for which a computational program was compiled.

The results of the numerical calculation provide a qualitative and quantitative assessment of the influence of suspended particles on the structure of a turbulent suspended flow. Based on the results of numerical calculations, the pattern of formation of the velocity field of the carrier fluid and suspended particles is described.

A solution is given to one of the complex problems of suspension-carrying flow hydraulics—calculation of the distribution of averaged velocity over the depth of the flow when the flow is unevenly saturated with suspension. Existing approaches to describing the distribution of speeds do not cover all available factors [5, 10, 40, 65, 79, 95] as a result, they mainly provide only a qualitative assessment of the nature of the distribution.

Thus, the proposed method for calculating the profile of average velocities of the carrier fluid and solid particles allows its use in further analysis of the processes of suspension and sediment transport as more universal, convenient, and reliable.

4.1.2 On the Average Velocities of Movement of the Phases of a Suspension-Carrying Flow

When studying the kinematics of suspension-carrying flows, it is of great scientific and practical interest to determine the value of the relative velocity of solid particles in relation to the velocity of the carrier fluid.

The averaged longitudinal component of the relative speed of movement is expressed by M. V. Velikanov [34] in the form:

$$\overline{w}_0 = \overline{u} - \overline{u}_s \tag{4.11}$$

where $\overline{u}, \overline{u}_s$—the averaged longitudinal component of the velocity of the carrier fluid and solid particles.

Formula (4.11) was subsequently used in their works by other authors [9, 27, 50, 84, 114, 144, 181].

Researchers pay special attention to determining the value of u_s. V. M. Makkaveev believes that $u_s > u$, that is, solid particles overtake liquid particles, since the main acting force is gravity, and the component of this force directed along the flow for heavy particles is greater than for water.

I. K., Orlov, and B. A. Fidman, on the contrary, indicate the lag of solid particles of the liquid. These conclusions are confirmed by experimental data [99].

The results of experimental studies of the turbulent movement of suspension-carrying flows allowed N. A. Mikhailova [109] to come to the conclusion that for smaller particles the difference in the velocities of liquid and solid matter will be significantly less.

Significant research in the field of hydrotransport has been carried out in this direction. It should be noted the works of M. A. Dementyev, N. A. Sililn, A. P. Yufin, V. I. Voitenko, V. M. Karasik, S. I. Kril, V. F. Ocheretko, M. V. Pechenkin, and other researchers.

Of interest are the conclusions made by M. A. Dementiev and M. V. Pechenkin [119], based on the results of experimental studies of the kinematics of suspension flows carried out on a special installation. The authors found that the averaged longitudinal velocities of the liquid phase in the main column of the flow are greater than the velocity of solid particles, and this difference decreases with increasing concentration.

Similar conclusions are presented in [9, 27, 144, 181], they indicate that the difference in the velocities of particles and liquid is observed on the dynamic axis. The magnitude of the relative longitudinal velocity is almost equal to the speed of free fall of particles in a fluid at rest.

It should be noted that at present there is no perfectly developed method for calculating relative speed. This is apparently due to the variety of factors characterizing the value of the relative velocity, such as size, density of suspended particles, etc.

We are considering the determination of the relative velocity value based on the averaged differential equation of motion (4.5). Solving the system of differential equations of motion will make it possible to obtain diagrams of the velocity distribution along the depth of the flow separately for the carrier fluid and separately for the flow of suspended particles. From here, using expression (4.11), we obtain the change in relative velocity along the depth of the flow.

The calculation results show that the value of the relative velocity changes along the depth of the flow. This is obvious, since when a suspension-carrying flow moves in open channels, the distribution of the concentration of suspended particles along the depth of the flow has an exponential character. Hence, with increasing depth, the concentration of suspended particles increases (see Chap. 5) and here it becomes necessary to take into account the fractional composition, which significantly affects the distribution of relative velocity along the depth of the flow (Fig. 4.1).

Calculations show that the relative longitudinal velocity is equal to the hydraulic size, only adjusted for constraint.

Figure 4.1 shows the dependence of the relative velocity on the flow saturation with suspension. And experimental data are plotted here [145]. As can be seen from the figure, as the concentration of sediment particles increases, the velocity of the solid particle lags far behind the flow.

Note that the obtained dependence for determining the relative velocity takes place for a certain size of sediment particles that the flow can carry in a suspended state.

For preliminary calculations, the expression for the relative speed is determined as follows: assuming a small concentration and taking $1 - s = 1$, the system of

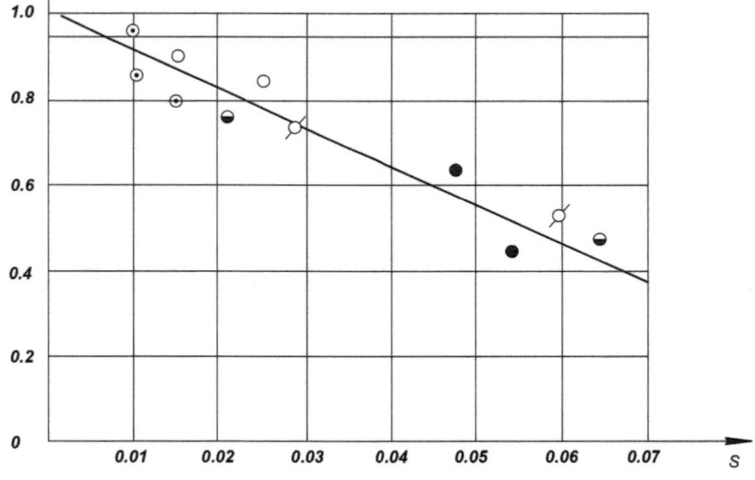

Fig. 4.1 Dependence of the relative difference in the speed of movement of particles and liquid on the concentration of suspension according to data from [139]

differential equations (4.5) from the second equation, we have:

$$u_s = \frac{Ku}{K + L_s s} + \frac{\rho_T g i s}{K + L_s s}$$

Taking into account (4.11), the formula for determining the longitudinal component of the averaged relative velocity of solid particles in relation to the velocity of the carrier fluid can be written as:

$$|u - u_s| = \left| \left(1 - \frac{K}{K + L_s s}\right) u - \frac{\rho_T g i s}{K + L_s s} \right|. \tag{4.12}$$

The calculation results show that for small fractions of suspended particles the difference in the velocities of the carrier liquid and solid particles is insignificant, which is confirmed by the conclusions presented in [1, 9, 33, 113].

Note that in this case the distribution of the average velocity over the depth of the flow is affected by the density and viscosity of the suspension-carrying flow (Chap. 3).

Based on the above, it can be noted that a solid particle cannot break away from the carrier liquid even in the case when $u_s > u$ the particle makes a vertical movement, thereby increasing the turbulence of the suspended flow.

4.2 Formation of the Velocity Field in Irrigation Canals

As noted above, velocity diagram profiles in open channels remain insufficiently studied even with a single-phase flow. Determining more refined velocity profiles requires the use of different hypotheses and approaches. The search for new directions in this aspect remains promising in solving the problem.

As mentioned above, in the process of studying the movement of fluid in various years, many formulas of an empirical or semi-empirical nature were proposed for calculating the velocity distribution over the flow cross section.

A detailed analysis of the theoretical premises of the velocity distribution formulas in the case of a single-phase flow is considered in [1, 2, 5, 7, 12, 13, 14, 17, 23, 25, 29, 35, 44, 50, 61, 62, 68, 86, 87, 96].

4.2.1 Velocity Distribution Along the Depth of Turbulent Flow in Irrigation Canals

Constructing a field of flow velocities requires knowledge of the pattern of changes in velocities in depth, or more precisely in some directions, which it is advisable to call velocity distribution lines. Under laminar flow conditions, where resistance to flow movement is determined only by the forces of molecular viscosity, the problem of velocity distribution for any shape of distribution lines is easily solved theoretically based on the equality between the shear force and the holding tangential force, proportional to the viscosity and the shear velocity gradient normal to the shear surface. For flows with a turbulent flow regime, where the main resistance to movement is created as a result of mixing water masses at equal speeds, there is no strict theoretical solution yet. Therefore, the problem of velocity distribution for turbulent flows is solved on the basis of semi-empirical theories or direct processing of experimental data.

To calculate the velocity distribution along the depth of the turbulent flow in irrigation canals, we proceed from Eq. (4.5).

In the case when single-phase flow $S = 0$, formula (4.5) is easily reduced to the form:

$$\mu \frac{\mathrm{d}^2 u}{\mathrm{d}z^2} - LU = \rho g i. \tag{4.13}$$

Solving Eq. (4.10), under the following boundary conditions:

$$z = 0; \qquad u = 0$$

$$z = H; \qquad \frac{\partial u}{\partial y} = 0.$$

We obtain the following expression for calculating the distribution of averaged velocity over the depth of a single-phase flow:

$$u = \frac{\rho g i}{L} \left(\frac{\mathrm{ch}\sqrt{\frac{L}{\mu}} \cdot z}{\mathrm{ch}\sqrt{\frac{L}{\mu}} \cdot h} - 1 \right) \tag{4.14}$$

where L is the turbulent exchange coefficient;

ch—hyperbolic cosine.

When fluid flows in pipelines and smooth surfaces, the value of L is given in works [5, 6, 57, 59]. As for open channels, the definition of the "L" value is much more complicated. Because this is related to the assessment of hydraulic resistance in channels, which are characterized by the mutual influence of the characteristics of the turbulent flow and the channel.

The first experimental studies in this direction are given in the works of Darcy (1850), Gangillier-Cutter (1869). The works of R. Manning (1889) are of particular importance in assessing channel resistance.

In recent years, when determining the roughness coefficient, the formula of A. Strickler has become widely known:

$$n = \left(\frac{0.15}{\sqrt{g}} \right) d^{\frac{1}{6}}$$

where d—average size of soil particles.

When determining the "*L*" *value in assessing the mutual influence of the flow and the channel*, the parameters of both the flow and the soil of the bed of the channel must be taken into account.

Following the above, based on the analysis of numerous experimental and field data from a number of scientists: Strickler, Manning, Gangillier, Kutter, Pavlovsky, etc., which were widely covered in the works [5, 6, 22, 23, 24, 57, 59, 79, 82, 84], we can draw the following conclusion.

By introducing the dimensionless parameter "La", defined as follows

$$\mathrm{La} = \frac{Lh^2}{\mu}.$$

When assessing the turbulent characteristics of the flow in [5, 6, 57, 59], La is determined as follows:

$$\mathrm{La} = 0.0025\mathrm{Re}$$

$\mathrm{Re} = \frac{\vartheta \cdot h}{\nu}$—Reynolds number.

Here the value ϑ is determined as follows:

Using the Darcy-Weisbach formula, we have:

$$\vartheta = \frac{\sqrt{8g}}{\lambda}\sqrt{Ri}.$$

Then, keeping in mind that the coefficient of hydraulic friction λ depends on the density, viscosity, speed, and roughness of the channel

$$\lambda = f(\rho, \mu, u, \Delta)$$

(where Δ is the height of the roughness protrusions), according to the theory of dimensions, we can obtain the following relationship between dimensionless quantities:

$$\lambda = f(\mathrm{Re},\ n)$$

or by entering the traditional value $\frac{C}{\sqrt{g}} = \frac{\sqrt{2}}{\lambda}$

($\frac{C}{\sqrt{g}}$—determined by the Strickler formula), we obtain:

$$\mathrm{La} = 0.0025\frac{\sqrt{i}}{n} \cdot \frac{\sqrt{g} \cdot h^{\frac{3}{2}}}{\nu}.$$

From here, using the dependence, $\mathrm{La} = \frac{Lh^2}{\mu}$ you can determine the value of the turbulent exchange coefficient "L".

$$L = 0.0025 \cdot \frac{\sqrt{i} \cdot \sqrt{g} \cdot \rho}{n \cdot h^{\frac{1}{2}}}$$

Thus, a new dependence is proposed to determine the value of "L", where both flow parameters and state are taken into account beds.

Putting the last expression in (4.1), we obtain a new dependence for calculating the velocity distribution along the flow depth, which takes into account, in addition to the turbulent characteristics of the flow, the influence of the channel bottom on the velocity distribution.

$$u = u_* \cdot n_0 \left(\frac{\mathrm{ch}\sqrt{\frac{L}{\mu}} \cdot z}{\mathrm{ch}\sqrt{\frac{L}{\mu}} \cdot h} - 1 \right) \tag{4.15}$$

where u_* is the dynamic speed;

n_0—reduced roughness coefficient.

4.3 Theory of Transport of Suspended Sediment Particles by Turbulent Flow

For a more complete description of the process of suspension, attraction, and transportation of river sediments, it is necessary to establish the nature of the distribution of suspended sediment particles along the depth of the flow.

The calculation formulas currently used in practice are mainly derived on the basis of diffusion theory. As noted above, this theory is based on the assumption that the coefficient of turbulent mixing of momentum and concentration is equal. The most successful expression turned out to be for determining ε

$$\varepsilon = \frac{\rho g i (1 - \eta)}{\frac{\mathrm{d}u}{\mathrm{d}y}}$$

However, determining the meaning $\frac{\mathrm{d}u}{\mathrm{d}y}$ poses some difficulties. Quite often, the value of this ratio is found empirically, so the resulting formulas for calculating the distribution of suspended sediment particles along the depth of the flow remain suitable only for specific cases. It is also $\frac{\mathrm{d}u}{\mathrm{d}y}$ determined independently of the influence of flow saturation with sediment.

A completely different approach is used when determining the distribution law for fluidized (boiling) flows [28, 45, 117, 149]. For example, N. I. Syromyatnikov [149], based on the convergence of the random movement of particles in a suspended layer under the influence of hydrodynamic flow forces and gravitational field forces with the thermal movement of gas molecules, as a result of experimental studies, proposed a calculation formula for the concentration distribution, reminiscent of the barometric formula of the molecular kinetic theory of gases

$$n = n_0 e^{-A_n h} \quad (4.16)$$

where n—number of particles at a certain height y;

n—number of particles at zero level;

A_n—coefficient proportional to the kinetic energy of moving particles.

B. A. Fidman [153, 154], applies dependence (4.13) for the case of distribution of solid particles in a liquid flow in a cylindrical pipe and thereby shows the possibility of using the molecular kinematic theory for the case of suspended flows.

Following what has been said [4, 20, 45, 117, 157], we consider the steady and uniform motion of the fluid. Let us assume that among the particles moving in the flow, there is a certain number of solid particles that move at a speed equal to the speed of the liquid. Note that this is confirmed based on field data [12].

Then, based on the equality of the kinetic energy of all particles, for an arbitrary fraction of suspended particles we can write the following equality

$$\frac{m_t \overline{u_s^2}}{2} = \frac{3}{2} K_0 T; \quad \frac{m_0 \overline{u^2}}{2} = \frac{3}{2} K_0 T,$$

where u, u_s are the speed of liquid and solid particles, respectively;

m_t—mass of solid particle;

m_0—mass of a solid particle that moves at a speed equal to the liquid;

K_0—Boltzmann constant;

T—temperature.

Hence the speed of a solid particle of an arbitrary fraction is equal to

$$\overline{u_s^2} = \left(\frac{d_0}{d_i}\right)^3 \overline{u^2},$$

where d_0—optimal diameter—the diameter of a solid particle that moves at a speed equal to the speed of the liquid.

Now let's look at determining the optimal diameter. According to the above model, i.e. when the speed of the solid particle is equal to the speed of the carrier fluid, the difference in the speeds of the vertical components is equal to zero. The speed of the carrier flow is expressed as follows:

$$u_x = u \cos \alpha$$
$$u_y = u \sin \alpha'$$

where α—angle of inclination of the flow to the horizontal;

u_x, u_y—respectively, the horizontal and vertical components of the velocity projection.

Each particle, being heavier than water and falling at a certain speed w relative to the mass of liquid surrounding it, will move vertically depending on the difference between the vertical component speed of the liquid and the speed of its own fall, then the speed difference will be determined by the expression

$$V_s = u_y - w,$$

where w—hydraulic fineness of suspended sediment particles.

In case $V_s = 0$, $u = u_s$ from here

$$u_y = w \tag{4.17}$$

The value of hydraulic fineness is determined by formula (3.55):

$$w = k\frac{g\rho_0 d^2}{4\nu} \tag{4.18}$$

where $p_0 = \frac{\rho_\mathrm{T} - \rho}{\rho}$;

ρ_T, ρ—respectively, the density of the solid particle and liquid.

Substituting (4.16) into (4.15), to express the optimal diameter we obtain the following relationship:

$$u \sin\alpha = k\frac{g\rho_0 d^2}{4\nu}.$$

From here

$$d_0 = \left(\frac{4\nu u \sin\alpha}{kg\rho_0}\right)^{1/2}. \tag{4.19}$$

Thus, we established the main parameters of the considered model.

Now let's look at the distribution of concentration over the depth of the flow. When particles are in a gravity field, the particle distribution changes under the influence of the Archimedes force. And the impulse experienced by solid particles n is equal to [157, 177]:

$\sigma = n\theta_0$,

where is $\theta_0 = \frac{2}{3} m_\mathrm{T} u_s^2$—a quantity that depends on the macroscopic parameters of the state of suspended particles [177].

$$\text{Momentum change } d\sigma = -g(\rho_\mathrm{T} - \rho) W_s n\, \mathrm{d}y,$$

where W_s—the volume is the expression $\sigma = n\theta$, and substituting the values θ we have

$$\begin{gathered} d\sigma = -g(\rho_\mathrm{T} - \rho) W_s \frac{\sigma}{\theta} \mathrm{d}y \\ d\sigma = -3g(\rho_\mathrm{T} - \rho)\frac{\sigma}{\rho_\mathrm{T} 2u_s^2}\mathrm{d}y. \end{gathered} \tag{4.20}$$

Integrating expression (4.18), under boundary conditions $\sigma = \sigma_0$, $y = H$ we obtain:

$$\sigma = \sigma_0 exp \left\{ -3g(\rho_T - \rho) \frac{1}{\rho_T 2u_s^2} (H - y) \right\}.$$

From here, moving on to the number of suspended particles, using the boundary conditions $n = n_0$ at $y = H$, the formula for determining the distribution of the concentration of suspended particles along the depth of the flow is:

$$s = s_0 \exp \left\{ -3g(\rho_T - \rho) \frac{1}{\rho_T 2u_s^2} (H - y) \right\}. \qquad (4.21)$$

Substituting (4.14) into (4.18), we write the formula for determining the distribution of an arbitrary fraction of suspended sediment particles along the flow depth:

$$s = s_0 \exp \left\{ -3g(\rho_T - \rho) \frac{1}{\rho_T 2u_s^2} \left(\frac{d_i}{d_0} \right)^3 (H - y) \right\}. \qquad (4.22)$$

Writing Eq. (4.18) as:

$$\varepsilon \frac{d\sigma}{dy} = -w\sigma,$$

where

$$\varepsilon = \frac{3g(\rho_T - \rho)}{2\rho_T w u_s^2},$$

and comparing with Eq. (1.4), it can be argued that the use of molecular kinetic theory made it possible, on the one hand, to compile a differential equation for the concentration profile, which in its form coincides with the known semi-empirical equation of turbulent diffusion, and on the other hand, to find an expression for the coefficient turbulent particle exchange.

Thus, formula (4.20) differs from existing ones in that, firstly, it was obtained using developments based on molecular kinetic theory, secondly, it takes into account a number of parameters characterizing the movement of suspended particles (etc.), thirdly d, p_m, p, C, through the optimal diameter, a number of characteristics of the carrier fluid are taken into account, which determine the nature of the suspension-carrying flows, which ultimately expands the scope of the calculation formula.

4.3.1 Method for Calculating the Distribution of Individual Sediment Fractions Along the Flow Depth

When developing engineering measures to combat harmful fractions of river sediment (especially when designing sedimentation tanks), it is necessary to calculate each sediment fraction. Retention of excessive amounts of sediment in settling basins, especially small fractions (silt particles), deprives irrigated areas of valuable mineral fertilizers and increases the volume of treatment work, and the passage of harmful fractions leads to blockage and siltation of irrigation canals. In this regard, a rational solution to this problem requires the correct determination of the amount of all sediment fractions entering irrigation canals.

A. N. Gostunsky, A. G. Khachatryan, A. V. Karaushev, K. I. Rossinsky, I. A. Kuzmin, Kh. Sh. Shapiro, Yu. A. Ibadzade and others studied this issue. Currently, there are three methods for calculating the distribution of river sediment size [89]:

(1) separation of sediment into separate fractions (fractional);
(2) description of the law of size change using empirical formulas;
(3) by the average size of the entire composition of sediments without taking into account the heterogeneity of their composition.

The first method of calculation is the most promising and generally accepted [9, 25, 79, 89, 112]. The third method gives large errors, and the mathematical description of the size distribution has not yet received a successful solution.

Following what has been said, we will consider the distribution of individual sediment fractions along the depth of the flow, based on the first method of fractional calculation (Fig. 4.2).

Using Eq. (4.20), we write the distribution of partial turbidity over the depth of the flow containing sediment of the ith fraction in the water as:

$$S_i = S_0 \exp\left\{ \frac{a}{u^2} \left(\frac{d_i}{d_0} \right)^3 (H - y) \right\}, \tag{4.23}$$

where $a = 3g \frac{(\rho_T - \rho)}{2\rho_T}$;

S_{0i}—turbidity of the ith fraction at the bottom.

The calculation of the distribution of total turbidity is performed by adding partial diagrams [79]:

$$S(\overline{y}) = \sum_{i=1}^{m_0} S_i(\overline{y}) \tag{4.24}$$

where is m_0—the number of all sediment fractions suspended in the flow.

When using formulas characterizing the exponential profile of the turbidity distribution over depth, there is difficulty in choosing the bottom boundary conditions, i.e. in determining bottom turbidity—S_0.

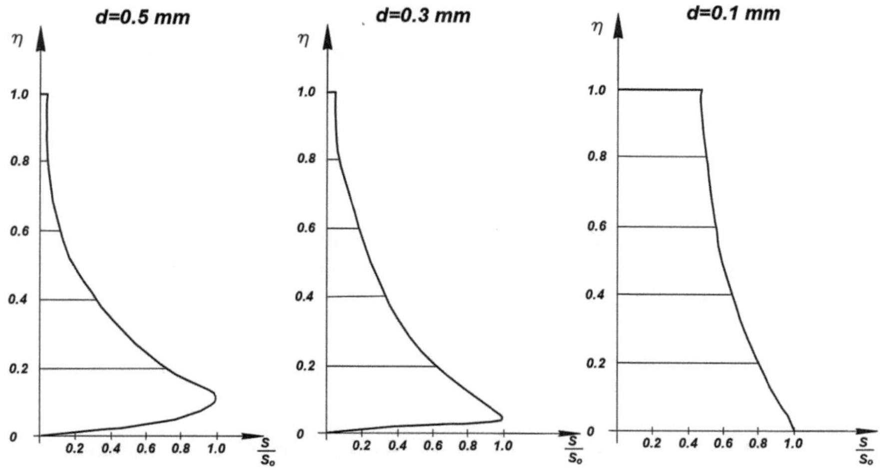

Fig. 4.2 Calculated characteristics of turbidity distribution along the flow depth depending on the diameter of suspended particles

Currently, various approaches are known to determine S_0. A number of researchers determine the value of bottom turbidity using empirical formulas [79, 188]. However, empirical formulas are suitable for special cases.

The authors of [143], based on experiments conducted to study the kinematics of suspension-carrying flows in a pipeline, claim that S_0 is observed at a distance of one or two solid particle sizes from the bottom of the pipe. V. N. Vinogradova [34], as a result of her research, also came to similar conclusions. The most complete idea on this matter was expressed by K. I. Rossinsky, V. K. Debolsky [140]: S_0 corresponds to the relative content of suspended sediment in the bottom region of the flow from $y = \Delta$ very small sediment and high flow velocities to $\Delta = h_m$ very large sediment. Some expressions for the definition Δ are given in [140, 141].

The works of K. I., R. Ossinsky, and I. A. Kuzmin indicate the possibility of expressing S_0 through S_{avg}. This approach greatly simplifies finding S_0 in contrast to the previous ones.

The replacement of S_0 by S_{av} was successfully used in the works of Kh. A. Irmukhamedov [72], Yu. A. Ibadzade [62], V. S. Sinelshchikov [146], and other researchers.

In the future, when using formula (4.20), the value of S_0 is expressed through S_{av}, which greatly simplifies the use of the formula in practical calculations.

Now let's consider the determination of S_{0i} based on the average turbidity value of a given fraction $S_{\text{av}\,i}$. Integrating expressions (4.20), we obtain the average vertical turbidity:

$$S_{cp_i} = \frac{S_{0_i}}{H} \int_0^H \exp(-\alpha_2(h - y))\mathrm{d}y, \tag{4.25}$$

where $\alpha_2 = \frac{\alpha}{V^2}\left(\frac{d_i}{d_0}\right)^3$.

After integration (4.23), we determine S_{0i} and substituting its value into (4.21), we get:

$$S_{0i} = \frac{\alpha_2 H S_{cpi}}{\left(1 - e^{\alpha_2 H}\right)} exp\{-\alpha_2(H - y)\}. \quad (4.26)$$

The value S_{cpi} can be obtained from hydrometric measurement data.

To determine the depth Y_i corresponding to the average turbidity value S_{cpi}, we make the following calculations:

$$S_{cpi} = S_{0i}\exp\{-\alpha_2(H - Y_i)\} \;\; S_{cpi} = \frac{S_{0i}}{\alpha_2 H}(1 - \exp\{-\alpha_2 H\}). \quad (4.27)$$

Solving the system of Eq. (4.25), we obtain the depth Y_i corresponding to the average turbidity value of the ith fraction on the vertical:

$$Y_i = H + \frac{1}{\alpha_2}\ln\left(\frac{(1 - \exp(-H\alpha_2))}{\alpha_2 H}\right). \quad (4.28)$$

Thus, two new dependencies (4.24) and (4.26) have been proposed to improve the application of formula (4.20) for calculation purposes. And formula (4.26) can be recommended for use in hydrometric measurements.

Field and Laboratory Studies of the Movement of a Suspended Flow

5

5.1 Brief Description of Research Objects

Parkent Main Canal (PMK). The canal passes through the territory of the Bostanlyk, Parkent, Yukori-Chirchik, and Akhangaran tumans of the Tashkent region. The canal is intended for irrigation of land on an area of 30.3 thousand hectares in the foothill zone of the Tashkent region. The canal belongs to class II structures, water intake is carried out from the Chirchik River.

Construction of the canal was carried out in two stages. The first stage, extending 69 km from PC 0 to PC 689 + 68, was built between 1975 and 1983. The second stage of the canal in the section from PC 689 + 68 (Kyzylsay) to PC 832 + 58 (Belyautsay), with a length of 14.3 km, was built in 1988–1997. The section from PC 832 + 58 to PC 856 + 10 was put into operation in 1998. The canal runs along the left bank of the river valley. Chirchik in difficult relief and geological conditions. In some areas, the excavation reaches 30–40 m, the embankment 20 m. Clayey, coarse-grained, semi-rocky and rocky rocks are common along the route. Clayey rocks are represented by heavy and loess-like loams, as well as clays. Coarse rocks-pebbles, boulders-pebbles, crushed stones, and gruss. Semi-rocky rocks are represented by siltstones, while rocky rocks are represented by limestones and syenites.

The operation of the canal and the structures on it is complicated by the fact that along the canal route there are highly subsiding soils, represented by loess-like loams.

The channel bed in the section PK 0-PK 341 + 70 and PK 873 + 17-PK 886 + 20 is made of concrete lining, the rest of the channel runs mainly in an earthen channel.

The canal is operated by the Parkent-Karasu Irrigation System Administration under the Chirchik-Akhangaran Basin Administration of Irrigation Systems (BAIS).

© The Author(s), under exclusive license to Springer Nature Singapore Pte Ltd. 2026

A. M. Arifjanov et al., *Dynamics of Suspended Flow in Open Channels*,
https://doi.org/10.1007/978-981-96-8781-7_5

Mirishkor Main Canal. The canal passes through the territory of Nishan, Muborak and Mirishkor districts of Kashkadarya region. The canal is intended for irrigation of land in the area 104,225 ha. The length of the channel is 118.9 km. The channel bed runs in an earthen channel and the cross-section of the channel is trapezoidal according to the design. Maximum flow rate 120 m^3/s. The length of the canal is divided into 5 sections and all sections run in an earthen channel. The purpose of this study was to study changes in the hydraulic elements of the canal, since after many years of operation the channel bed changes and takes on a more or less natural stable shape.

Field studies on the Parkent and Mirishkor canals were carried out during the period from 2006 to 2009.

Based on the results of field data, the pattern of formation of the velocity field is described, as well as the solution to one of the complex problems of hydraulics—the calculation of the distribution of averaged velocity over the depth of the flow.

Karakum Canal. The bed of the Karakum Canal passes through easily eroded soils. The changing nature of the Amu Darya and the difficulties in operating the damless water intake determined the multi-head nature of the Karakum Canal, consisting of four head sections.

The choice of the supply channel as a site for field measurements was due to the fact that, firstly, the Karakum Canal originates from the Amu Darya and is fed by almost the most turbid water in Eurasia (in certain periods, the turbidity in the river reaches 15–20 kg/m^3), secondly, due to the damless water intake from the river, the regime of solid flow is not regulated, thirdly, the site of field measurements is in many ways similar to the conditions of the Amu Darya [77], and this makes it possible to compare the data obtained with other formulas derived for the conditions of the river.

Thus, field studies of the hydraulic characteristics of the flow allowed us to collect data that allowed us to make a quantitative assessment of the proposed methods for calculating the distribution of average velocity over the depth of the flow, the distribution of turbidity over the depth and length of the flow.

5.2 Analysis of Field Research Data

An assessment of the distribution of turbidity along the depth of the flow based on field data on the Karakum Canal, carried out by analyzing turbidity diagrams, showed that the saturation of the flow with suspended sediment mainly increases towards the bottom and only 12.7% of cases correspond to a decrease in turbidity towards the bottom, and 2% is the greatest turbidity at a point of 0.6 N. As a rule, the highest turbidity of the flow corresponds to the highest average flow velocity, but at the bottom, with a turbidity of the flow of more than 8 kg/m^3, it is almost impossible to record the speed; the signals coming from the rotation of the turntable are inconsistent and rare, which indicates irregular flow, large pulsations. The speed measured at such points with a simple pinwheel does not correspond to reality. For example, at a point near the bottom on the nearest vertical to the left bank on June 24, a turbidity of 2.4 kg/m^3 was measured, and a speed was recorded at 0.48 m/s, and with a turbidity of 11.9 kg/m^3, the speed was determined to be 0.25 m/s. Table 5.1 shows the size of suspended sediments of the Karakum Canal.

The size of suspended sediment from the Mudarya River in the study area is given in Table 5.2. Sediment size varies significantly throughout the year. Table 5.2 provides data for the Mudarya River and the Karakum Canal, broken down into three main fractions: the largest and smallest particle size values along the channel cross-section in each month are selected.

The relative content of particles varies within the following limits for the Mudarya region: $d > 0.05$ mm—15.64–74.97%, particles $d = 0.05$–0.01 mm—13.70–58.18%, particle $d < 0.01$ mm—6.35–33.30%, i.e. the content of particles in the flow $d = 0.05$–0.01 mm is subject to the smallest fluctuations, but this value, equal to 4, is still significant. In general, it can be seen that during high water the percentage of particles $d < 0.01$ mm in the flow increases, and during low water the content of large particles increases.

Data on the size of bottom sediments obtained from analyzes of samples taken at the research site indicate that the main channel-forming sediments are sand particles 0.5–0.1 mm (Table 5.3).

The specific gravity of bottom sediments is a fairly stable value and is in the range of 2.62–2.88 t/m^3, and the volumetric weight is 1.37–1.49 t/m^3.

The head part of the Karakum Canal consists of 4 head canals, through which water is taken from the Mudarya River and supplied, somewhat clarified in settling tanks, to regulatory structures: along the third head to the new one with a flow rate of 350 m^3/s, to the old one with a flow rate of 170 m^3/s along the second head and sluice. The length of the first settling tank (downstream) is 3.4 km; in recent years, this head of the canal has been inactive. The length of the second settling tank of the Amu Darya is carried out most often through two canal heads—II and III.

Water flow rates in the canal varied within wide limits: in the second head from 20 to 220 m^3/s, in the third head from 179 to 350 m^3/s. The minimum measured flow rate in head II was 108 m^3/s, and in head III—227 m^3/s, the remaining flow

Table 5.1 Coarseness of suspended sediments in the study area

Month	Extreme values	Fraction composition, %			D_{avg}, mm
		0.05 mm	0.05–0.01	0.01 mm	
May	Max	53.20	20.10	26.70	0.055
	Minimum	30.36	26.04	43.60	0.035
July	Max	65.76	22.46	11.79	0.182
	Minimum	13.69	55.01	31.30	0.03
October	Max	74.07	16.80	9.10	0.17
	Minimum	24.90	27.08	48.02	0.03

Table 5.2 Coarseness of suspended sediments of the Amu Darya and Karakum Canal

Month	Sampling location	Extreme values	Fraction composition, %			d_{avg}, mm
			$d > 0.05$ mm	0.05–0.01 mm	d < 0.01 mm	
May	Amu Darya	Max	35.85	31.6	33.6	0.052
		Min	34.8	25.30	28.30	0.038
	Channel (head)	Max	53.20	20.10	26.70	0.055
		Min	33.36	26.04	43.60	0.035
July	Amu Darya	Max	45.94	33.40	20.66	0.099
		Min	15.64	52.45	31.91	0.030
	Channel (head)	Max	65.75	22.46	11.79	0.182
		Min	13.69	55.01	31.33	0.030
	Downstream of the regulator	Max	52.64	31.36	16.00	0.050
		Min	19.94	49.42	30.64	0.030
October	Amu Darya	Max	60.08	28.05	11.87	0.130
		Min	59.67	25.96	14.37	0.060
	Channel (head)	Max	74.07	16.80	9.10	0.170
		Min	24.90	27.08	48.02	0.030
	Downstream of the regulator	Max	64.92	24.20	10.88	0.059
		Min	27.60	41.30	31.10	0.036

Table 5.3 Fractional composition of bottom sediments of the Mudarya River

Month	Particle diameter and content by weight				
	1.05–0.5	0.5–0.25	0.25–0.1	0.1–0.05	0.05–0.01
May	0.04	62.0	31.6	2.5	3.5
June	0.07	60.0	34.8	3.0	0.5
July	14.0	78.0	5.0	0.5	2.5
September	6.0	50.4	40.0	1.5	2.0

elements at minimum flow rates for head II and head III were $h_{av} = 2.8$, $h_{max} = 4.98$ m, $v_{av} = 1.1$ m/s, $v_{max} = 1.63$ m/s.

The river flow entered the head of the channel quite smoothly, and only behind the upper left projections in both heads II and III, whirlpool zones were formed, the width of which reached 1/6 of the width of the channel of head III. The velocities measured here were small, but the depths were significant. The maximum measured depth in the canal was 5.95 m with a water flow of 584 m m^3/s in the month of June. The composition of suspended sediments at the head of the channel is given in Table 5.3.

Studies have shown that the distribution of linear flow rates of water and sediment is more uniform at the head of the canal than in the river. The average turbidity on the verticals in the channel differs slightly from each other, and when compared with the distribution of sediment along the verticals in the river, it turns out that in the river bed, sediment is distributed more evenly in depth, which is possibly explained by the fact that in the channel the turbulence of the flow in depth is more uniformly developed.

When comparing the average flow turbidities calculated for the entire cross section of the river and the canal (up to the settling tanks) with simultaneous measurements, it turns out that the saturation of the flow with sediment in the canal is greater than in the river, despite the fact that the water velocities in the canal are lower than in the river (Table 5.4). The turbidity of the flows is levelled out only when the ratio of the water speed in the river v_{p} to the speed in the channel v_k is equal to 1.36. However, studies have shown that when designing connecting channels or settling tanks in calculations, it is not entirely correct to take the average turbidity in the channel based on the average turbidity of the river calculated for the entire cross-sectional flow. The average turbidity of the flow at the head of the canal is influenced by the turbidity of those streams of the river flow that flows near the water intake, i.e. in the zone of its influence on the river flow, and it was found that the zone of influence varied from 1/6 V at flow rates in the river more than 4000 m^3/s to 1/3 V—at flow rates less than 1500 m^3/s (B is the width of the river along the water's edge). It would be more correct to use the maximum turbidity value in the calculations based on the unfavourable (in the sense of turbidity distribution across the river cross-section) flow in the water intake area. As studies have shown, the uneven distribution of turbidity in the river section under consideration was 2–6.

The fractional composition of suspended sediments in the river and at the head of the canal up to the settling tanks is identical. The fractional composition of sediments both in the river and in the canal varies widely. Particles $d > 0.05$ mm are subject to the greatest fluctuations, and $d = 0.05–0.01$ mm are subject to the least fluctuations. During high water, the average particle size in the river and in the canal is smaller than during low water (Fig. 5.1).

Table 5.4 Fractional analysis of suspended sediments of the Karakum Canal

Study area	d, mm	At the bottom		0.8 N		0.6 N		0.2 N	
		%	$\sum$%	%	$\sum$%	%	$\sum$%	%	$\sum$%
Karakum Canal	0.001	2.92	2.92	3.58	3.58	7.2	7.2	5.27	5.27
	0.005	3.46	6.36	5.44	9.02	3.05	10.25	9.48	14.75
	0.01	2.61	8.99	3.33	12.35	4.29	14.54	5.48	20.23
	0.05	22.25	31.24	34.44	46.79	35.85	50.39	46.64	66.87
	0.1	46.6	77.84	44.29	91.08	43.49	93.88	30.32	97.19
	0.25	10.97	88.81	5.55	96.63	3.74	97.62	2.03	99.22
	0.5	11.19	100	3.37	100	2.38	100	0.78	100

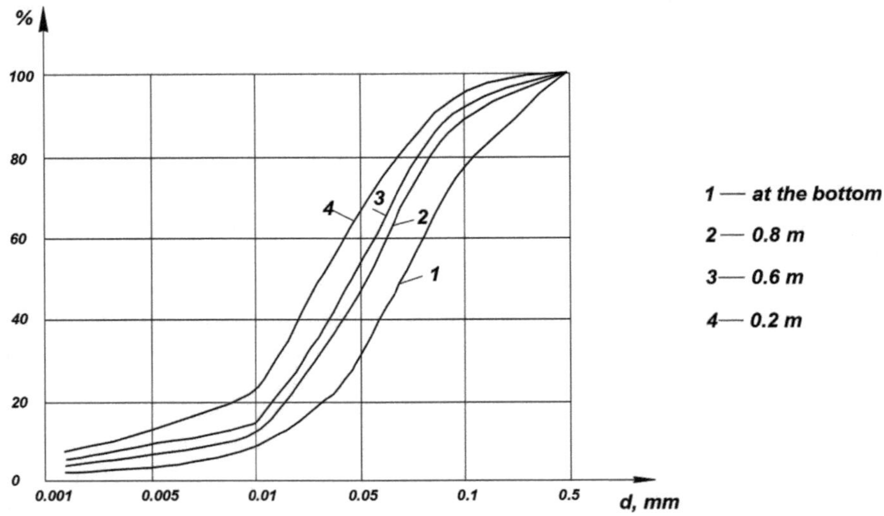

Fig. 5.1 Fractional analysis curves of suspended sediments of the Karakum Canal

5.2.1 Change in Stream Turbidity

Let's consider the issue of turbidity distribution along settling tanks, across the cross-section and in depth.

Hydraulic flow elements at study sites, Karakum Canal

Month	No	ϱ, m^3/c	B, m	ω, m^2	h_{cp}, m	h_{max}, m	v_{cp}, m /s	v_{max}, m/s	S, kg/m^3
June	1	310.01	137	389.9	2.85	5.4	0.77	0.91	1.86
	2	331.6	100	272.9	2.73	4.0	1.21	1.58	1.73
	3	273.9	132	275.5	2.09	3.8	0.99	1.37	1.66
	4	237.5	90	237.5	2.64	3.7	1.0	1.19	1.33
July	1	310.2	129	257.8	2.00	3.4	1.24	1.31	3.22
	2	310.05	110	321.8	2.92	4.34	0.96	1.41	2.0
	3	259.10	135	371.25	2.75	5.45	0.70	0.79	1.92
	4	265.12	98	364.55	3.72	5.16	0.72	0.88	1.59
	5	317.08	132.5	366.55	2.72	4.05	0.86	1.05	2.27
	6	256.4	72.5	263.5	3.63	4.6	0.97	1.09	1.64
August	1	245.0	135	214.95	1.57	3.0	1.14	1.86	2.64
	2	344.4	104	253.76	2.44	3.45	1.35	1.48	2.10
	3	290.1	124	303.8	2.45	4.65	0.95	1.36	3.2
	4	190.2	86	203.0	2.36	3.75	0.94	1.50	2.69

(continued)

(continued)

Month	No	Q, m^3/c	B, m	ω, m^2	h_{cp}, m	h_{max}, m	v_{cp}, m /s	v_{max}, m/s	S, kg/m^3
	5	336.6	141	346.9	2.46	3.7	0.97	1.17	2.23
	6	178.5	88	203.3	2.31	4.11	0.88	1.17	2.42
September	1	200.8	125.5	251.3	2.0	3.7	0.8	1.12	1.01
	2	199.1	108	259.2	2.4	4.45	0.73	1.04	0.72
	3	190.2	135	253.95	1.88	2.95	0.75	0.93	0.92
	4	189.4	86	218.4	2.54	3.55	0.86	1.08	0.87
	5	193.4	121	215.38	1.78	2.95	0.90	1.32	0.96
	6	184.6	83	173.8	2.09	2.75	1.06	1.27	0.95
October	1	221.45	125.5	265.45	2.11	3.45	0.83	1.09	1.05
	2	176.9	104	247.15	2.38	3.60	0.72	1.23	1.02
	3	225.0	130	256.26	1.97	3.1	1.00	1.03	1.1
	4	135.7	80	200.0	2.50	3.5	0.68	1.09	0.98

The measurements showed that the distribution of turbidity along the depth of the flow in the overwhelming majority is of a well-known nature—it increases from the surface to the bottom, and the distribution profile resembles a “boot”. The highest measured turbidity at a point near the bottom was 10.85 kg/m^3, and the bottom velocity was 1.36 m/s.

5.3 Analysis of the Fractional Composition of Suspended Sediment Particles

We will dwell on the analysis of the fractional composition of suspended sediment particles in more detail, since the nature of the turbidity distribution along the depth of the flow largely depends on the size of the suspended sediment particles. Taking this factor into account contributes to a more complete description of the concentration field of suspended sediment particles.

Table 5.4 shows the results of the fractional composition of suspended sediment particles of the Karakum Canal, broken down into seven fractions. As can be seen from the analysis of the fractional composition of suspended sediment particles, suspended sediment consists of all particles composing the channel bed only in different percentages.

An analysis of the fractional composition of suspended sediment particles, depending on the flow regime, shows that there are certain dependencies on the flow regime; they show that there are a certain amount of suspended sediment particles that do not change depending on the flow regime. This conclusion once again shows the correctness of the approach chosen in the formation of the law, the distribution of turbidity along the depth of the flow. Now let’s look at the analysis of the turbidity distribution diagram along the flow depth based on field measurements. A characteristic feature of the turbidity distribution diagrams along

Table 5.5 Fractional analysis of suspended sediments of the Mirishkor channel

Study area	*d* (mm), % yes							
	> 0.25	0.25–0.10	0.10–0.05	0.05–0.01	0.01–0.005	0.005–0.001	< 0.001	Physical clay
PC-620	0.0	1.4	24.7	65.3	2.5	4.0	2.1	8.6
PC-720	0.0	0.6	34.3	55.9	3.8	3.1	2.3	9.2
PC-933	0.0	0.6	40.5	53.6	3.1	1.6	0.6	5.3
P K-1060	0.0	0.9	27.1	60.4	7.1	2.4	2.1	11.6

the depth of the flow is the appearance of a "boot". The results of the analysis show that this is due to the presence of large fractions of suspended sediment particles in the suspended flow.

As the results of fractional analysis of suspended particles show, in the case when the fraction of particles of a certain size ($d > 0.1$ mm) predominates in the fractional composition, the distribution diagram takes on the appearance of a "boot".

As can be seen from the analysis results, in June, particles $d = 0.1$ mm make up almost 82% of the composition of suspended particles, and in October, particles $d = 0.1$–0.05 mm make up 59% of the composition of suspended sediment particles.

Accordingly, the turbidity distribution diagrams along the flow depth have different shapes (Table 5.4).

The proportion of flow turbidity with particles $d > 0.1$ mm in June is 0.75 kg/m^3, and in October 0.32 kg/m^3. This is two times less than in June.

This is explained by the fact that during low-water periods the suspension capacity of the flow for particles of larger fractions is less. As a result, large fractions cannot be involved in the thickness of the flow and they will move in the bottom layer of the flow, and the amount of particle $d < 0.01$ mm remains almost constant. As for the Mirishkor canal, there is a different picture (Table 5.5). In the composition of suspended particles, the amount of particle $d > 0.25$ mm is practically absent. This is due to the fact that large particles are retained in settling basins (Table 5.6).

5.4 Analysis of the Method for Calculating the Distribution of Averaged Longitudinal Velocity over the Depth of a Suspended Flow

In order to check the reliability of the obtained dependencies for calculating the distribution of average velocity along the depth of the suspended flow, as well as for a comparative analysis of existing methods for calculating the distribution of velocity along the depth of the flow, we used mainly field data obtained on the Karakum Canal, and also processed some results of field measurements carried out on Karshi Canal and in the lower reaches of the Amu Darya River, in

Table 5.6 Fractional analysis of suspended sediments of the Karakum Canal under different flow regimes

No.	*d*, mm	0.2 N		0.6 N		0.8 N		At the bottom	
		%	$\sum$%	%	$\sum$%	%	$\sum$%	%	$\sum$%
For the month of June (v = 1.5 m/s)									
1	0.001	10.43	10.43	15.64	15.64	7.37	7.37	3.6	3.6
2	0.005	18.11	28.54	5.8	21.44	7.42	14.79	3.79	7.39
3	0.01	9.39	37.93	5.93	27.37	2.32	17.11	3.3	10.69
4	0.05	51.2	89.13	39.15	66.52	42	59.11	28.5	39.19
5	0.1	10.4	99.53	33.04	99.56	40.1	99.21	59.74	98.93
6	0.25	0.3	99.83	0.29	99.85	0.52	99.73	0.73	99.66
7	0.5	0.17	100	0.15	100	0.27	100	0.34	100
For the month of October (v = 1.0 m/s)									
1	0.001	10	10	9.21	9.21	5.45	5.46	5.03	5.03
2	0.005	2	12	0.48	9.69	10.91	16.47	0.77	5.8
3	0.01	13.6	25.6	3.27	12.96	11.62	28.19	0.66	6.46
4	0.05	37.2	62.8	25.34	38.3	29.2	57.39	13.3	18.76
5	0.1	33.12	95.92	54.44	92.74	31.12	88.51	61.54	80.3
6	0.25	3.59	99.51	6.21	98.95	10.05	98.56	16.85	97.15
7	0.5	0.49	100	1.02	100	1.54	100	1.85	100

which the author took part. As already noted, to calculate the distribution of averaged velocities over the flow depth, there are various empirical semi-empirical dependencies obtained under different conditions and modes. We used formulas (4.1)–(4.4) as a comparative analysis of the proposed formula (4.5) with other existing dependencies. This is due to the fact that the indicated formulas were obtained for the conditions of p.A. Mudarya, and field data were collected under the same conditions and close to them. The results of a comparison of field data and calculated data (4.5) are given in Table 5.7. Figure 5.2 shows a comparative analysis of formula (4.5) with the formulas of elliptic, exponential, parabolic, and logarithmic profiles. As can be seen from Fig. 5.2, the comparison results indicate the suitability of using formula (4.5) for calculation purposes.

Field studies of the movement of the suspension-carrying flow on the Karakum Canal allowed us to collect data that allows us to make a quantitative assessment of the formulas for determining the distribution of the average velocity along the depth of the suspension-carrying flow.

Figure 5.3 shows the results of calculating the distribution of the average velocity over the depth of the suspended flow using formula (4.5), where the values of the measured data on the Karakum Canal are also plotted. The calculation results are summarized in Table 5.16, which also shows the percentage of discrepancy between measured and calculated data.

Table 5.7 Results of comparative analysis

Study area	I, %	Depth	Y/H	Measured speed, U, m/s	Design speed				
					Parabolic	By logarithmic	According to elliptical	According to indicative	According to formula (4.5)
Karakum Canal	0.015	1.26	At POV	1.00	1.00	1.0	1.0	1.00	1.00
			0.2H	0.98	0.940	0.998	0.98	0.98	0.99
			0.6H	0.89	0.849	0.995	0.869	0.87	0.90
			0.8H	0.82	0.797	0.986	0.783	0.79	0.84
			At the bottom	0.75	0.68	0.974	0.69	0.74	0.755
	0.015	1.46	At POV	0.98	0.98	0.98	0.98	0.98	0.98
			0.2H	0.96	0.966	0.977	0.968	0.95	0.97
			0.6H	0.93	0.926	0.975	0.871	0.921	0.94
			0.8H	0.89	0.765	0.964	0.77	0.82	0.92
			At the bottom	0.84	0.64	0.951	0.635	0.70	0.795

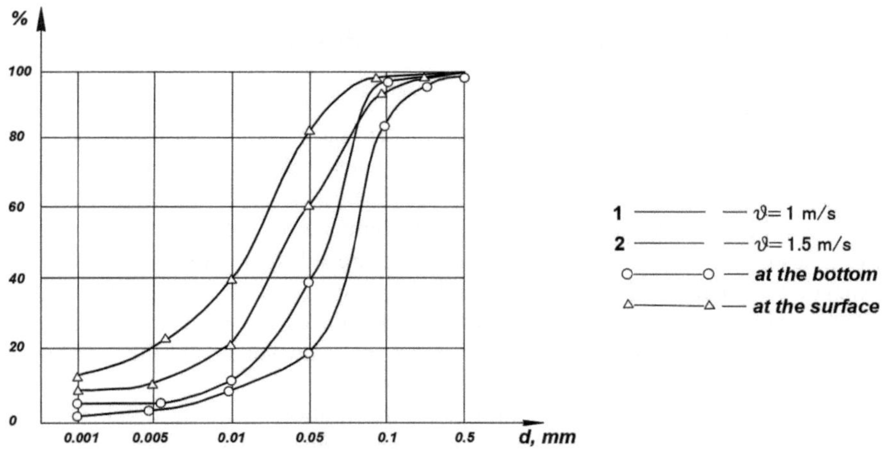

Fig. 5.2 Fractional analysis curves of suspended sediments under different movement modes

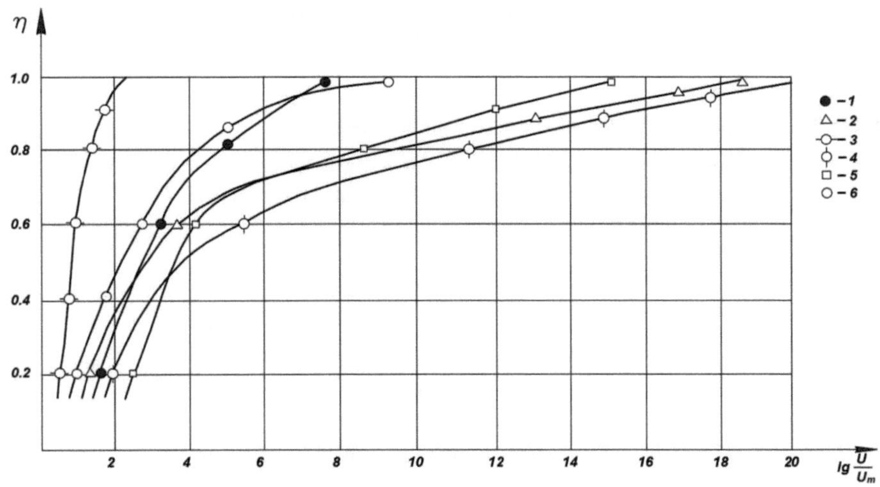

Fig. 5.3 Comparison of calculated data obtained using the formulas of parabolic (2), logarithmic (3), elliptical (4), exponential (5) profiles and the formula proposed by the author (6)

As the results of the analysis show, the proposed method for calculating the distribution of the average velocity over the depth of a suspended flow can be used with sufficient accuracy for calculation purposes.

5.4.1 The Influence of the Concentration of Suspended Sediment Particles on the Average Velocity Distribution

A qualitative assessment of the influence of the concentration of suspended particles on the distribution of average speed was carried out by a number of authors [2, 5, 72, 78, 112, 115].

However, it is very difficult to quantify the influence of suspended matter on the dynamics of the suspended flow. Currently, the prevailing opinion is that energy is expended to maintain solid heavy particles in suspension. The work of weighing takes away the energy of the flow, that is, it reduces the intensity of turbulent mixing. The consequence of this is an increase in the unevenness of the distribution of averaged velocities over the depth of the flow.

In the above works, the calculation of the distribution of averaged velocity over the depth of a suspended flow is carried out using formulas derived for single-phase flows (4.1–4.4) with clarification of empirical coefficients $m, æ, x, p$ (Table 5.8).

For the middle and lower reaches of the Amudarya River, the values of the coefficient ($m, æ, P, x$) were studied by A. M. Mukhamedov [115] and for these conditions the values x and P were taken constant, equal to $\chi = 0.12$; $P = 0.58$. Offered m and x determine depending on the Chezy coefficient. Whereas in work

Table 5.8 Comparison of calculation results with field data

Study area	Depth H, m	Surface speed max., m/s	Slope I, %	Relative depth, η	Measured speed, U	Design speed, U	Differences %
Karakum channel	4.7	1.0	0.015	POV	1.0	1.0	0.0
				0.8	0.98	0.96	2.0
				0.6	0.84	0.82	2.3
				0.4	0.80	0.76	5.0
				At the bottom	0.42	0.44	4.7
	4.46	0.98	0.015	POV	0.98	0.98	0.0
				0.8	0.98	0.97	2.0
				0.6	0.87	0.90	– 3.4
				0.4	0.80	0.82	– 2.5
				At the bottom	0.62	0.60	3.2
	2.8	0.81	0.015	POV	0.81	0.81	0.0
				0.8	0.80	0.79	2.0
				0.6	0.80	0.78	– 3.0
				0.4	0.74	0.72	– 2.8
				At the bottom	0.5	0.45	– 4.0

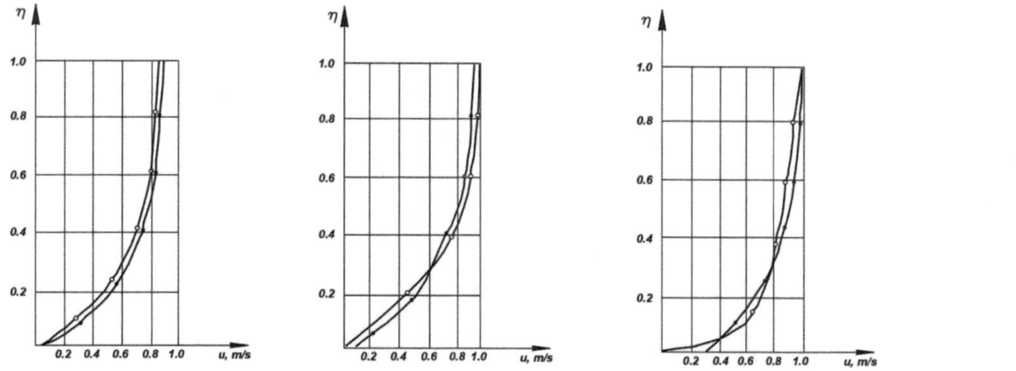

Fig. 5.4 Comparison of calculation results using formula (4.5) with field data measured on the Karakum Canal 0-0—measured; --- calculated

[71] the following data are presented, where the value of the coefficients χ, P varies within the range: $P = 0.28–1.0$; $\chi = 0.06–0.56$.

The variability of these coefficients indicates that through these coefficients it is not possible to take into account the uneven distribution of the concentration of suspended particles along the depth of the flow.

In the case of a suspended flow, when the flow consists of a carrier liquid and a solid particle, the structure of the flow has its own characteristics. Here, the nature of the velocity distribution is influenced not only by the saturation of the flow with suspension, but also by the size of the suspended particles, the mutual influence of the carrier fluid and solid particles, the depth distribution of the latter, etc. (Fig. 5.4).

Field studies carried out on the Karakum Canal show that the uneven distribution of turbidity along the depth of the flow significantly affects the distribution of average velocity. Here we can note such an interesting fact that with almost identical flow parameters and terrain conditions, but with different saturations of the flow with suspended matter, different velocity distribution profiles along the depth of the flow are already obtained (site 1, June, August).

In Table 5.9 shows the results of calculating the distribution of the average velocity over the depth of the suspension-carrying flow according to formula (4.5) at various saturations of the flow with suspension, where the values of the measured data on the Karakum Canal are also plotted (Fig. 5.5).

5.4.2 Analysis of the Method for Calculating the Distribution of Averaged Velocity over the Depth of a Single-Phase Flow

Now let's consider the possibilities of using formula (4.12) to calculate the distribution of averaged velocity over the depth of a single-phase flow.

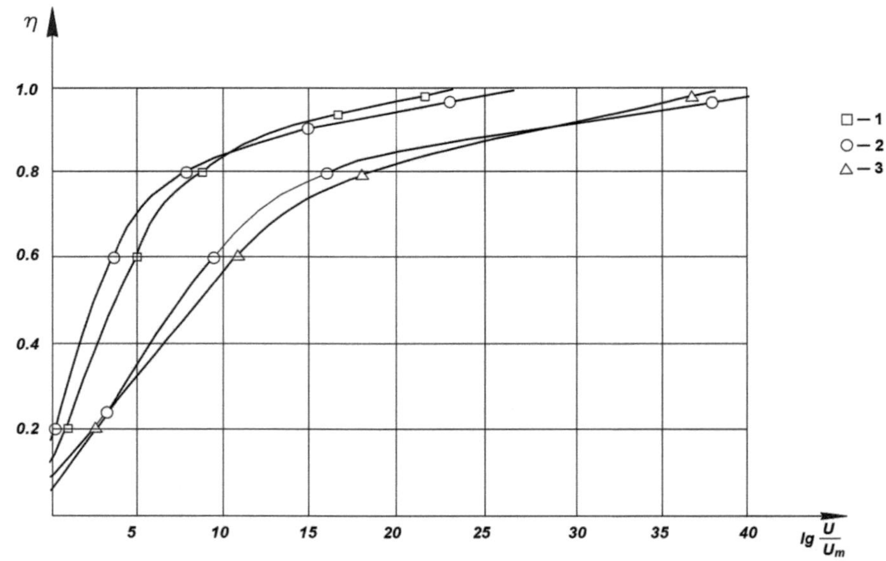

Fig. 5.5 Profiles of distribution of averaged velocity over flow depth 1—$S = 0.03$; 2—$S = 0.01$; 1, 2—measured data; 3—calculated data

Table 5.9 Velocity distribution over depth at different flow saturations with suspended matter

Study area	Depth H, m	Suspension concentration	Slope I, %	Relative depth η	Measured speed, 10^{-2} $\lg u/u_{\max}$	Design speed, 10^{-2} $\lg u/u_{\max}$	Differences %
Karakum Canal	2.2	0.03	0.015	POV	–	–	–
				0.8	2.7	2.72	2.0
				0.6	10.8	9.7	2.5
				0.4	18.7	15.6	7.0
				At the bottom	39.8	42.0	4.0
	2.06	0.01	0.015	POV	–	–	–
				0.8	1.6	0.87	3.0
				0.6	5.06	4.6	2.0
				0.4	9.7	8.6	2.5
				At the bottom	22.2	23.7	4.0

h, the slope of the water surface i, the roughness coefficient n and the average diameter of soil particles d are taken.

For known h, I, n using formula (4.12), we calculate the distribution of velocities over the depth of the flow. The following hydraulic elements were obtained from naturalistic studies: $h = 3.0$ m, $i = 0.00006$, $n = 0.0225$.

$$u = u_* \cdot n_0 \left(\frac{ch\sqrt{\frac{L}{\mu}} \cdot y}{ch\sqrt{\frac{L}{\mu}} \cdot h} - 1 \right)$$

where is u_* the dynamic speed;

$$u_* = \sqrt{ghi} = \sqrt{9.81 \cdot 3.0 \cdot 0.00006} = 0.04\, M/c$$

n_0—reduced roughness coefficient;

$$n_0 = \frac{n}{0.0025} = 9;$$

L—coefficient turbulence, we will determine based on formula (4.11);

$$L = 0.0025 \cdot \frac{\sqrt{i} \cdot \sqrt{g} \cdot \rho}{n \cdot h^{\frac{1}{2}}} = 0.0025 \frac{\sqrt{0.00006 \cdot 9.81} \cdot 1000}{0.0225 \cdot \sqrt{3}} = 1.57$$

μ—coefficient of dynamic viscosity, $\mu = 0.015 \frac{Hc}{m}$;
y—coordinate axis; $y = \eta \cdot h$;
η—relative depth.

Asking _ a series of values, we determine the flow velocity on the vertical, that is, on the surface, at depths of 0.2, 0.4, 0.6, 0.8 N and at the bottom.

The calculation can be done on a computer using the well-known simple EXSEL program.

Based on the calculation results, we construct a velocity distribution profile along the depth of the flow (Figs. 5.6, 5.7, 5.8 and 5.9; Tables 5.10, 5.11 and 5.12).

To check the possibility of using the derived equation for calculation purposes, laboratory data were also used (laboratory studies were carried out under the guidance of Candidate of Technical Sciences, senior researcher M. R. Karapetyan).

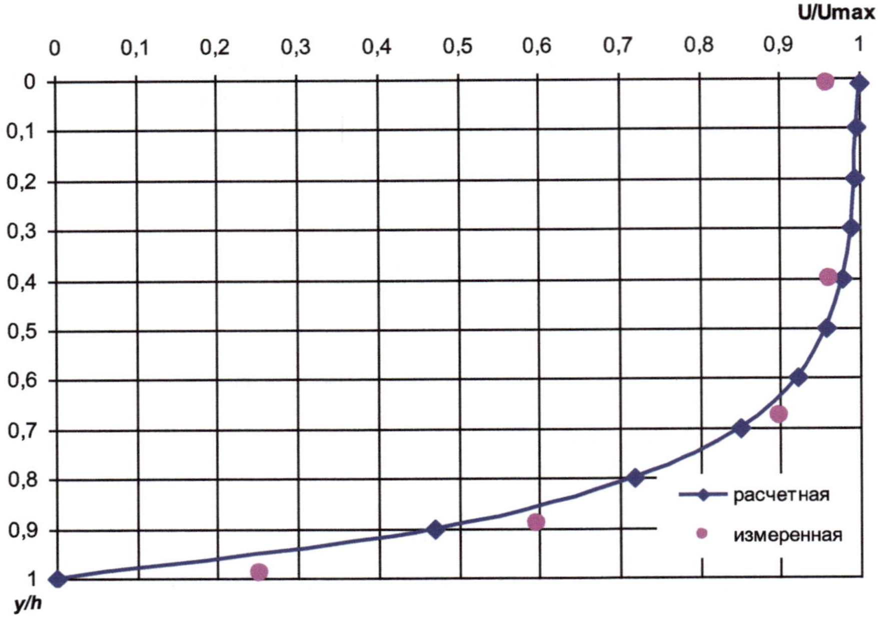

Fig. 5.6 Velocity distribution along the depth of the flow according to the proposed method

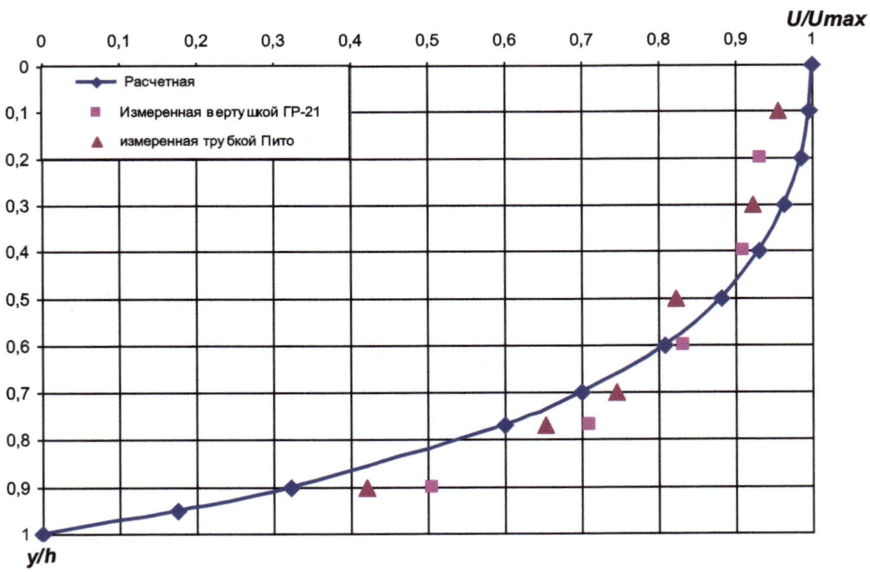

Fig. 5.7 Results of field studies of velocity distribution along the flow depth on the Parkent Canal (PK475+00)

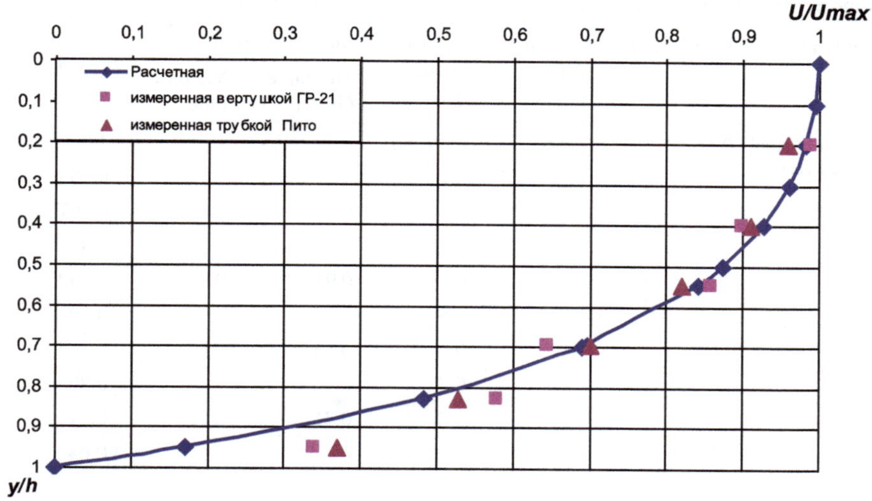

Fig. 5.8 Results of field studies of velocity distributions along the flow depth on the Parkent Canal (PK 540+00)

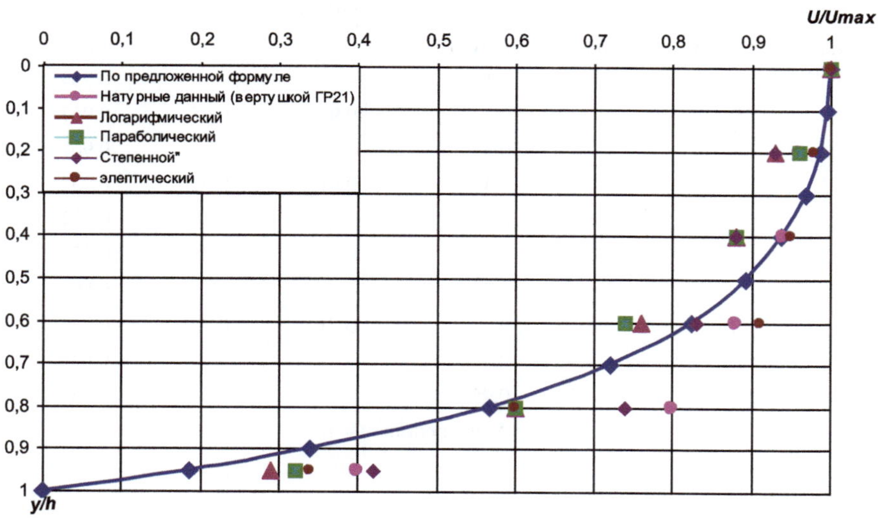

Fig. 5.9 Comparison of calculated and measured data

The results of a comparison of measured and calculated data are summarized in Table 5.13.

Table 5.10 Velocity distribution along the flow depth on the Parkent Canal

Study area	Depth h, m	Bottom slope, i	Relative depth η	Measured speed, u/u_{max}		Design speed, u/u_{max}
				Turntable GR-21	Via pitot tube	
Parkent Canal PC-475+00	2.56	0.000115	POV (0)	–	–	1.0
			0.1	–	–	0.989
			0.2	0.93	0.95	0.967
			0.3	–	–	0.967
			0.4	0.91	0.92	0.932
			0.5	–	–	0.891
			0.6	0.83	0.82	0.857
			0.7	–	–	0.805
			0.8	0.71	0.72	0.741
			0.9	–	–	0.642

Table 5.11 Velocity distribution along the flow depth on the Parkent Canal

Study area	Depth h, m	Bottom slope i	Relative depth η	Measured speed, u/u_{max}		Design speed, u/u_{max}
				Turntable GR-21	Via pitot tube	
Parkent Canal PC-540+00	2.45	0.000115	POV (0)	–	–	1.0
			0.1	–	–	0.989
			0.2	0.93	0.95	0.967
			0.3	–	–	0.967
			0.4	0.91	0.92	0.932
			0.5	–	–	0.891
			0.6	0.83	0.82	0.857
			0.7	–	–	0.805
			0.8	0.71	0.72	0.741
			0.9	–	–	0.642
			0.95	–	–	0.383
			At the bottom (1.0)	–	–	0

As the results of the analysis show, the proposed method for calculating (4.12) the distribution of average velocities over the flow depth can be used for calculation purposes with sufficient accuracy and versatility.

Table 5.12 Results of a comparative analysis of the calculation of velocity distribution over flow depth

Depth *h*, m	Bottom slope, *I*	Relative depth η	Measured speed, u/u_{max}	Design speed, u/u_{max}				Design speed, u/u_{max}
				Logarithmic	Along a parabolic	By power	Elliptical	
4.92	0.00002	POV (0)	–	–	–	–	–	1.0
		0.2	0.96	0.93	0.96	0.93	0.98	0.98
		0.4	0.94	0.88	0.88	0.88	0.95	0.93
		0.6	0.88	0.76	0.74	0.83	0.91	0.82
		0.8	0.8	0.6	0.6	0.74	0.6	0.56
		At the bottom (1.0)	–	–	–	–	–	0
5.92	0.00002	POV (0)	–	–	–	–	–	1.0
		0.2	0.99	0.96	0.98	0.96	0.98	0.99
		0.4	0.94	0.93	0.93	0.93	0.95	0.95
		0.6	0.85	0.86	0.86	0.88	0.89	0.85
		0.8	0.7	0.73	0.73	0.8	0.78	0.68
		At the bottom (1.0)	–	–	–	–	–	0

5.5 Analysis of the Method for Calculating Turbidity Distribution over Flow Depth

Determining the nature of the distribution of turbidity along the flow depth is one of the main links in studies of river sediments. There are various methods and approaches to solving this problem, which were analyzed in detail in previous chapters.

As was said, we considered this issue on the basis of molecular kinetic theory and obtained Eq. (4.20) for calculating the distribution of turbidity over the depth of the flow:

$$S = S_0 \exp\left\{-\left(\frac{a}{u^2}\left(\frac{d_i}{d_0}\right)^3\right)(H-y)\right\}, \quad 0 < y < H \tag{5.1}$$

where $a = \frac{3g(\rho_m - \rho)}{2\rho_m}$.

Now let's consider the possibilities of using this formula for calculation purposes. The analysis of this formula was carried out by comparing and contrasting the calculated data with field and laboratory data from a number of researchers, as well as the author. Here, data from laboratory studies by N. A. Mikhailova [109] and the results of field measurements by G. I. Shamov [169] in the lower reaches of the river were used. Amu Darya; B. K. Balakaev [25] on the Karakum Canal.

Table 5.13 Comparison of calculation results using formula (4.5) with laboratory data

Depth H, m	Surface speed, m/s	Slope, 10^{-3}	Relative depth	Measured speed	Design speed	Discrepancies
0.77	0.277	0.01	POV	0.227	0.227	0.0
			0.8	0.226	0.225	1.9
			0.6	0.224	0.22	1.7
			0.4	0.220	0.20	– 9.0
			0.2	0.185	0.18	2.7
			At the bottom	0.17	0.15	9.0
0.061	0.215	0.01	POV	0.215	0.215	0.0
			0.8	0.215	0.21	1.8
			0.6	0.20	0.20	0.0
			0.4	0.18	0.17	5.5
			0.2	0.145	0.16	– 10.3
			At the bottom	0.12	0.18	– 8.3
0.091	0.228	0.01	POV	0.228	0.228	0.0
			0.8	0.22	0.22	0.0
			0.6	0.21	0.21	0.0
			0.4	0.20	0.20	0.0
			0.2	0.18	0.19	5.5
			At the bottom	0.14	0.15	– 7.1

Table 5.14 shows a comparison of the calculation results using formula (5.1) with the data of G. I. Shamov, and with the data of B. K. Balakaev, the discrepancy between the calculation results using formula (5.1) and the measured data is on average 8–10%. The results of comparison with the formulas of I. F. Karasev and A. V. Karaushev are also given here.

For comparative analysis, the formulas of M. V. Velikanov and A. V. Karaushev were proposed, since, according to a number of researchers, these formulas more convincingly describe the distribution of turbidity along the depth of the flow.

A comparative analysis of formula (5.1) with the calculation formulas of M. V. Velikanov and A. V. Karaushev, as well as with natural data, shows the advantage and suitability of the proposed formula (5.1) for calculation purposes. The calculation results are summarized in Table 5.15 (Fig. 5.10).

Table 5.14 Comparison of the calculation results using formula (5.1) with the field data of G. I. Shamov

Plot research	*N*, m	*V*, m/s	*w*, m/s	Δ*H*	*η*	Turbidity in kg/m^3		Differences %
						S measured	*S* calculated	
R. Amu Darya	4.35	1.50	0.0008	0.30	POV	1.0	0.79	+ 21
				0.87	0.2	1.20	0.97	+ 21
				2.61	0.6	1.75	1.90	− 8.5
				3.48	0.8	1.45	2.70	− 10.2
				4.35	Bottom	3.85	3.85	0.0
				0.30	POV	1.0	1.05	− 5
R. Amu Darya	3.62	2.25	0.0008	0.72	POV	1.12	1.16	− 3.5
				2.20	0.2	1.25	1.47	− 17.6
				2.9	0.6	1.56	1.66	− 6.4
				3.62	0.8	1.88	1.88	0.0
				0.20	Bottom	1.20	1.30	− 8.3
R. Amu Darya	2.0	1.85	0.0008	0.32	0.2	1.25	1.40	− 12.0
				0.97	0.6	1.62	1.70	− 4.9
				1.30	0.8	1.88	1.87	0.5
				1.62	Bottom	2.06	2.06	0.0

Yu. A. Ibadzade [65], comparing Vanoni's experiments with calculations using various formulas (Table 5.16), concludes that the formulas of N.V. Karaushev and V. M. Makkaveev are acceptable for calculation purposes. Comparison of our formula (5.1) with Vanoni's data gives identical results (Fig. 5.11).

It is known that the most detailed laboratory studies of the movement of suspended particles in a liquid flow were carried out by N. A. Mikhailova [109]. A comparison of the specified data with the results of calculation using formula (5.1) is given in Table 5.17. As the results of the analysis show, the measured and calculated data provide satisfactory convergence.

This is apparently due to the fact that formula (5.1) takes into account a number of factors (*d*, *C*, *K*, *w*) that characterize the behaviour of a solid particle in a liquid flow. Taking into account these parameters contributed to qualitatively, in a new form, describing the distribution of the concentration of suspended particles along the depth of the flow. And the comparison results provided a quantitative assessment of the proposed model for the distribution of suspended particles (Fig. 5.12).

Table 5.15 Comparison of calculation results using formula (5.1) with natural data by B. K. Balakaev

Place of selection samples	Average		η	Turbidity, kg/m^3								
				Measured		Calculated by formulas						
	Depth H, m/s	Speed V, m/s		S	S/S_0	I. F. Karaseva	A. B. Karausheva		Yu. A. Ibadzade		(5.1)	
							S	S/S_0	S	S/S_0	S	S/S_0
Pk-5	1.95	1.01	0.03	3.058	1.00	1.00	3.67	1.00	3.025	1.00	3.099	1.00
			0.20	3.023	0.99	0.935	3.359	0.919	2.825	0.935	2.918	0.96
			0.60	2.3676	0.875	0.812	2.997	0.817	2.493	0.824	2.785	0.91
			0.80	2.386	0.783	0.760	2.842	0.775	2.375	0.787	2.537	0.81
			1.00	2.207	0.724	0.710	2.763	0.753	2.207	0.696	2.42	0.70
Pk-10	1.80	0.79	0.03	2.71	1.00	1.00	3.059	1.00	2.678	1.00	2.686	1.00
			0.20	2.64	0.975	0.915	2.77	0.901	2.486	0.93	2.528	0.941
			0.60	2.25	0.830	0.770	2.42	0.789	2.183	0.814	2.356	0.87
			0.80	1.998	0.736	0.706	2.335	0.759	2.091	0.781	2.046	0.762
			1.00	1.77	0.653	0.644	2.293	0.746	1.77	0.66	1.900	0.71
Pk-15	2.80	0.80	0.03	2.126	1.00	1.00	2.556	1.00	1.95	1.00	2.106	1.00
			0.20	2.086	0.98	0.913	2.242	0.876	0.896	0.973	1.91	0.907
			0.60	1.695	0.796	0.768	1.961	0.776	1.598	0.82	1.717	0.815
			0.80	1.543	0.725	0.703	1.864	0.730	1.487	0.761	1.390	0.658
			1.00	1.35	0.624	0.637	1.802	0.704	1.352	0.693	1.25	0.591

Table 5.16 Results of comparative analysis

Study area	Depth H, m	Average speed U, avg. m/s	Hydraulic size ω, m/s	Slope I, %	Relative depth $\eta = Y/H$	Turbidity, kg/m^3			
						Measured	Calculated by formula		
							M. V. Velikanova	A. Karausheva	(5.1)
Karakum Canal	2.4	0.65	0.00195	0.00015	0.2	0.64	0.3	0.55	0.61
					0.6	0.7	0.4	0.62	0.69
					2.8	0.82	0.52	0.78	0.86
					At the bottom	1.0	1.0	1.0	1.0
	2.4	0.68	0.00008	0.00015	0.2	0.59	0.88	0.97	0.47
					0.6	0.64	0.91	0.97	0.58
					2.8	0.73	0.94	0.98	0.82
					At the bottom	1.0	1.0	1.0	1.0

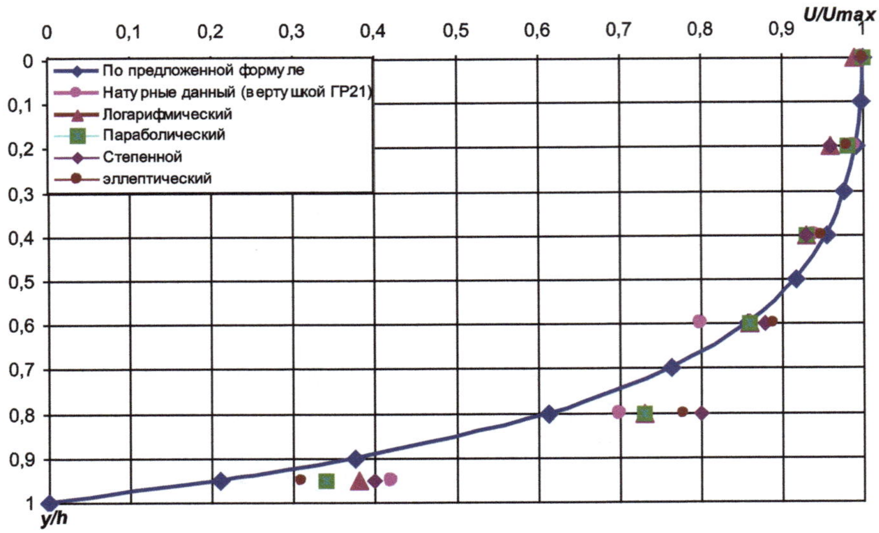

Fig. 5.10 Comparison of calculated and measured data

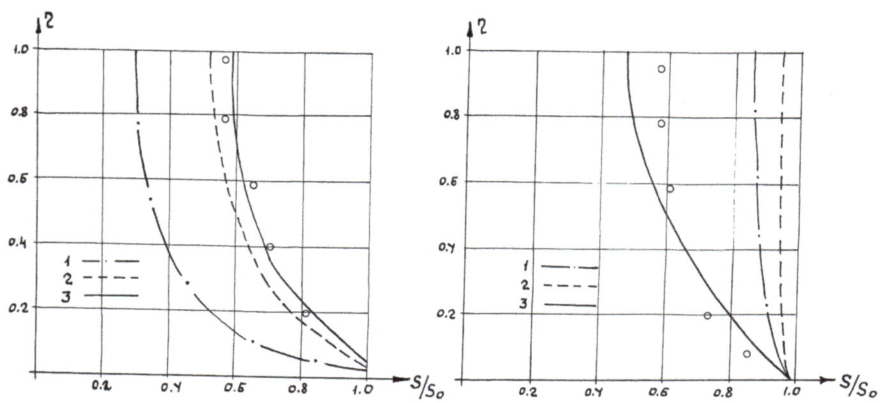

Fig. 5.11 Comparison of calculated data with field data. 1—according to Velikanov's formula, 2—according to Karaushev's formula, 3—according to the author's formula, 0-0—natural data

5.5.1 Comparison of the Results of Calculating the Distribution of Individual Sediment Fractions Along the Flow Depth

As is known, river sediments are always heterogeneous. Therefore, there is a need to take into account the different fractions of sediment in the calculations. As noted above (Chap. 4), currently there are various ways to take into account the different

Table 5.17 Results of comparison of calculations using formula (5.1) with Vanoni's experimental data

N, m	U, m/s	ω, m/s	Distance from the bottom, m	Results of determination of turbidity using formulas of various authors						
				Makkabeeva	Velikanova	Karausheva	Greshaeva	Taylor	Karmana	(5.1)
1	0.98	0.008	0	100	100	100	100	–	–	100
			0.1	86	37	86	58.6	86	86	87.6
			0.2	73	29	75	47.8	60.5	71	77.21
			0.4	53.8	22.1	58.2	34.6	41.5	53	59.61
			0.6	38.6	14.1	46.1	26.6	35.2	34.8	46.0
			0.8	28.9	5.9	38.1	20.9	24	17.5	35.54
			1	29.9	0	31	16.9	16.9	0	28.53

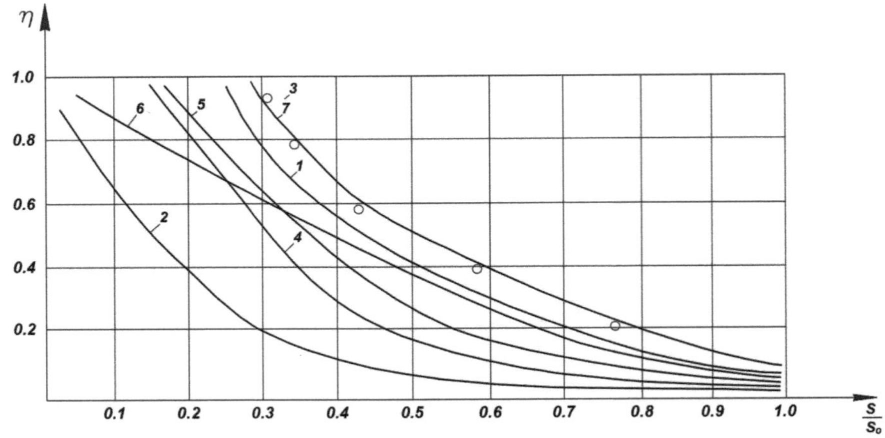

Fig. 5.12 Comparison of calculated data with Vanoni's experiments using the formulas: 1—Makkaev; 2—Velikanova; 3—Karausheva; 4—Grishaeva; 5—Taylor; 6—Pocket; 7—author; 0—experienced data

fractionalities of suspended particles in calculations. In the description of the turbulence of a suspended flow, it was noted (Chap. 4) that particles whose diameter is less than 0.01 have a speed equal to the speed of the carrier fluid and hence their distribution pattern along the flow depth differs from the pattern of distribution of larger suspended particles. The nature of the distribution of suspended particles with a diameter greater than 0.01 mm is close to the exponential law of distribution of each fraction of river sediment.

Figure 5.13 provides a calculation of the distribution of individual sediment fractions along the depth of the flow, where field data measured on the Amudarya River and in various sections of channels are also plotted. The calculation results are summarized in Table 5.17.

Note that above (Chap. 4), for the convenience of choosing boundary conditions, a calculation dependence was proposed for calculating the value of the average turbidity on the vertical using formula (4.28).

Figure 5.14 presents the results of a comparison of the calculated and measured average turbidity values on the vertical. The calculation results are summarized in Table 5.18. The comparison results indicate a satisfactory convergence of measured and calculated data (Figs. 5.15 and 5.16; Tables 5.19, 5.20, 5.21, 5.22 and 5.23).

Thus, the results of comparison and comparison of the calculation formula (5.1) indicate the possibility of its use for solving practical problems of determining the distribution of turbidity along the depth of the flow, including taking into account the different fractionalities of river sediments.

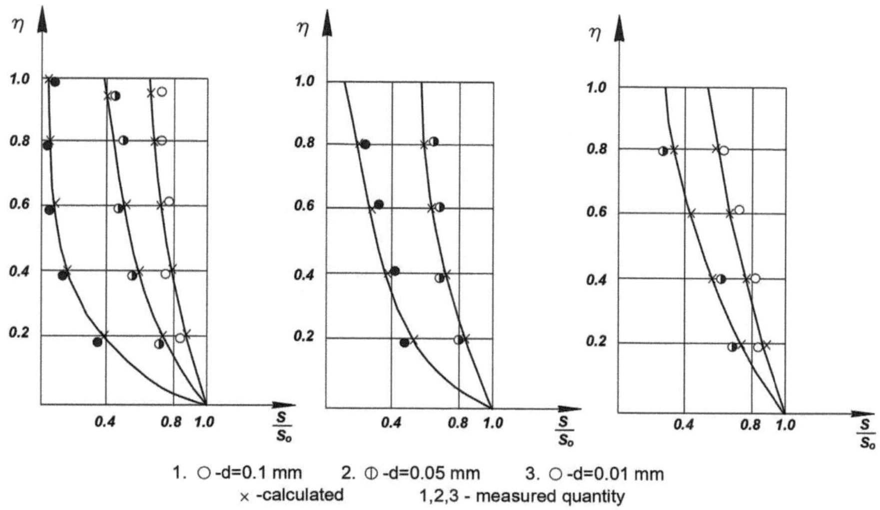

Fig. 5.13 Comparison of natural data (a-p. Amu Darya; Karakum Canal; β-channel Kizketgen) and calculations using formula (5.1) for particles of different fractions

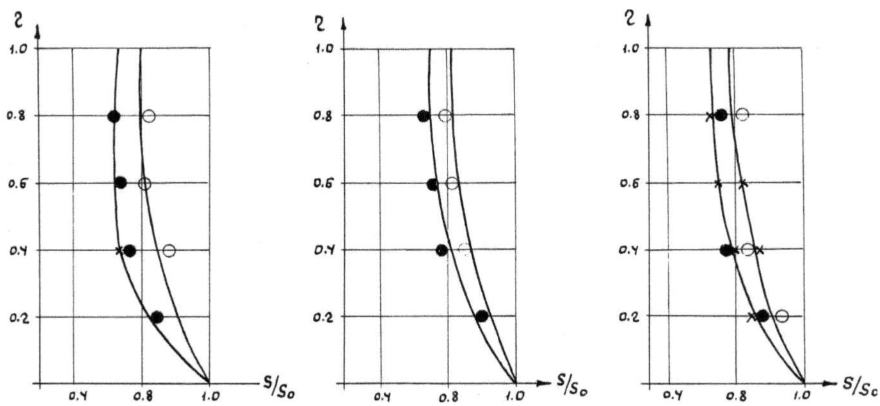

Fig. 5.14 Results of comparison of measured and calculated data for particles of different fractions. 1.0-0 $d = 0.05$ mm, 2.0-0 $d = 0.01$ mm

Table 5.18 Comparison of the calculation results using formula (5.1) with the experimental data of N. A. Mikhailova (measured with an Orlov bathometer)

No.	No. of experiments	η	Turbidity		No.	No. of experiments	η	Turbidity		No.	No. of experiments	η	Turbidity	
			S calc. kg/m^3	*S* change kg/m^3				*S* calc. kg/m^3	*S* meas. kg/m^3				*S* calc. kg/m^3	*S* meas. kg/m^3
1	13	0.096	0.36	0.36	2	31	0.059	0.182	0.182	3	36	0.092	0.0154	0.054
		0.138	0.27	0.41			0.077	0.157	0.187			0.116	0.86	0.906
		0.180	0.185	0.33			0.092	0.152	0.170			0.140	0.78	0.450
		0.284	0.126	0.248			0.281	0.106	0.0929			0.438	0.23	0.45
		0.328	0.107	0.115			0.307	0.101	0.098			0.474	0.20	0.082
		0.422	0.075	0.096			0.522	0.6070	0.063			0.811	0.051	0.031
		0.658	0.032	0.015			0.555	0.063	0.056			0.836	0.46	0.037
		0.842	0.016	0.20								0.061	0.041	0.040

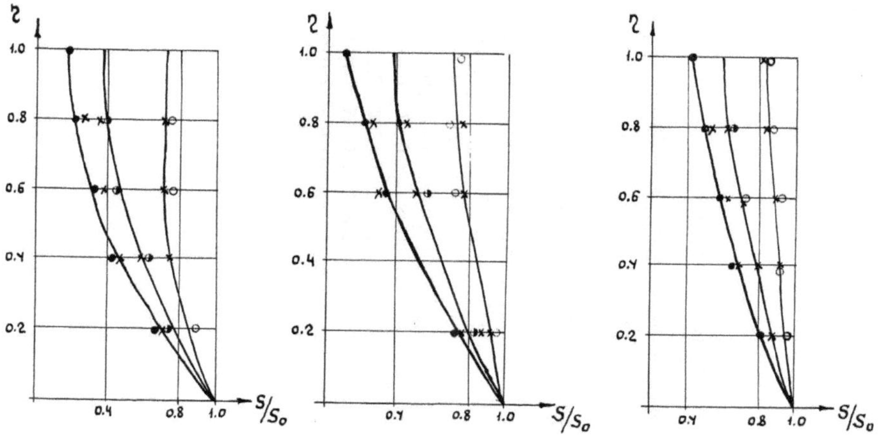

Fig. 5.15 Results of comparison of measured and calculated data for particles of different fractions. 1.0-0 $d = 0.05$ mm, 2.0-0 $d = 0.01$ mm, 3.0-0 $d = 0.005$ mm

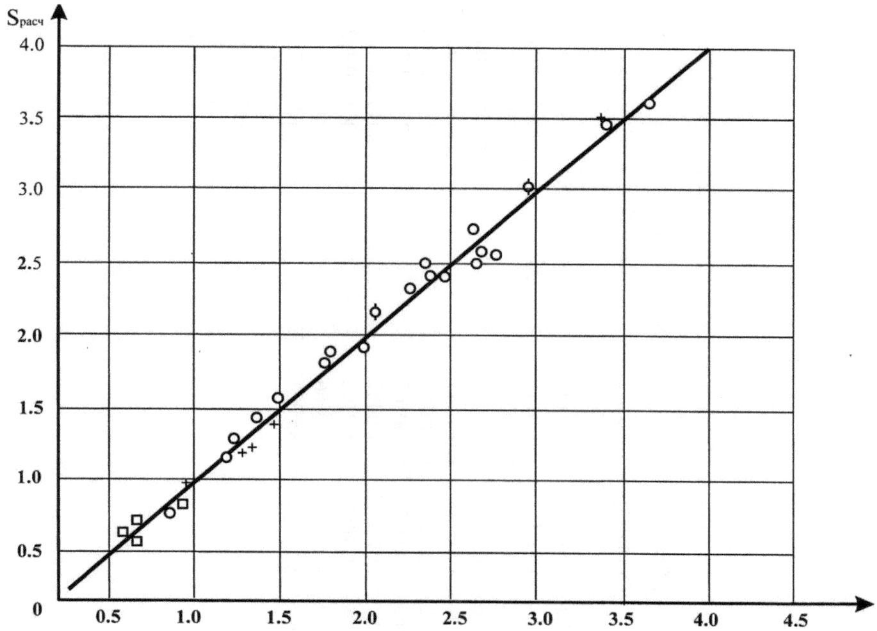

Fig. 5.16 Comparison of field data with calculation results using the attached formula: 1, 2—Karakum channel; 3—r.Amu Darya; 4—Karshi channel

Table 5.19 Compilation of calculation results using formula (4.20) with experimental data from N.A. Mikhailov (measured by filming)

No.	No. of experiments	η	Turbidity		No.	No. of experiments	η	Turbidity		No.	No. of experiments	η	Turbidity	
			S calc. kg/m^3	*S* meas. kg/m^3				*S* calc. kg/m^3	*S* meas. kg/m^3				*S* calc. kg/m^3	*S* meas. kg/m^3
1	32	0.044	0.000427	0.00042	3	35	0.69	0.021	0.00221	5	49	0.05	1.00	1.00
2	34	0.064	0.000127	0.000141	4	38	0.98	0.00102	0.00141			0.083	0.529	0.633
		0.084	0.00008	0.000088			0.126	0.00084	0.000067			0.150	0.388	0.250
		0.086	0.00105	0.00105			0.155	0.00068	0.000335			0.217	0.254	0.143
		0.144	0.000317	0.000843			0.184	0.00055	0.000107			0.283	0.168	0.096
		0.142	0.000236	0.000464			0.213	0.0004	0.000456			0.350	0.110	0.057
		0.169	0.000178	0.000167			0.027	1.37	1.428			0.483	0.0476	0.0348
		0.197	0.000133	0.000087			0.045	1.07	1.14			0.583	0.0254	0.018
		0.225	0.000099	0.000106			0.063	0.422	0.78			0.683	0.0135	0.0132
		0.253	0.000073	0.000066			0.081	0.33	0.359			0.717	0.0109	0.0108
		0.306	0.0000042	0.0000043			0.099	0.258	0.195					
							0.117	0.201	0.0813					
							0.135	0.157	0.425					
							0.153	0.123	0.011					
							0.171	0.096	0.005					

Table 5.20 Comparison of calculation results using formula (5.1) with full-scale data by G. I. Shamov, K. I. Baymanov and the author

Plot research	N, m	U, m/s	Depth relative $\eta = Y/H$	Fraction weight in kg/m^3					
				Measured			Calculated		
				$d = 0.1$ mm	$d = 0.05$ mm	$d = 0.01$ mm	$d = 0.1$ mm	$d = 0.05$ mm	$d = 0.1$ mm
R. Amu Darya	4.5	1.50	0.2	0.025	0.44	0.75	0.02	0.4	0.68
			0.4	0.04	0.56	0.83	0.039	0.5	0.71
			0.6	0.18	0.69	0.47	0.18	0.69	0.84
			0.8	0.39	0.83	0.92	0.42	0.7	0.92
			At the bottom	1.0	1.0	1.0	1.0	1.0	1.0
Karakum channel	2.4	0.77	0.2	0.13	0.64	0.59	0.12	0.61	0.47
			0.6	0.29	0.95	0.64	0.28	0.69	0.64
			0.8	0.54	0.83	0.68	0.5	0.85	0.9
			At the bottom	1.0	1.0	1.0	1.0	1.0	1.0
Kyzketgen channel	3.92	1.39	POV		0.22	0.56		0.73	0.56
			0.2		0.33	0.72		0.87	0.62
			0.6		0.75	0.82		0.86	0.78
			0.8		0.92	0.84		0.93	0.88
			At the bottom		1.0	1.0		1.0	1.0

Table 5.21 Results of comparison of the calculation formula with the data of I. I. Goroshkov

Place of selection samples	Depth N, m	Speed V, m/s	η	Fraction weight in kg/m^3			
				Measured		Calculated	
				0.05 mm	0.01 mm	0.05 mm	0.01 mm
Channel Klych-bai	2.40	0.80	0.2	0.7	0.84	0.7	0.73
			0.6	0.79	0.88	0.79	0.85
			0.8	1.0	1.0	1.0	1.0
	2.20	1.25	0.2	0.74	0.8	0.74	0.77
			0.6	0.79	0.88	0.97	0.87
			0.8	1.0	1.0	1.0	1.0
	2.00	1.00	0.2	0.77	0.68	0.73	0.77
			0.6	0.86	0.9	0.85	0.86
			0.8	1.0	1.0	1.0	1.0

Table 5.22 Results of comparison of calculations using formula (5.1) with field data /21/

No.	Date and site of study	N, m	U, m/s	η	Fraction weight in kg/m^3					
					Measured			Calculated		
					0.005 mm	0.02 mm	0.005 mm	0.05 mm	0.02 mm	0.005 mm
1	Kyz-ketgen channel 1978 P K 20	3.92	1.39	POV	0.22	0.56	0.073	0.27	0.56	0.8
				0.2	0.33	0.72	0.87	0.36	0.64	0.84
				0.6	0.75	0.82	0.86	0.59	0.78	0.91
				0.8	0.92	0.89	0.93	0.92	0.88	0.95
				At the bottom	1.0	1.0	1.0	1.0	1.0	1.0
2	Kegaili channel 1978 GK125	3.25	1.19	POV	0.16	0.37	0.7	0.18	0.5	0.71
				0.2	0.37	0.63	0.67	0.25	0.6	0.74
				0.6	0.52	0.76	0.75	0.5	0.77	0.86
				0.8	0.79	0.81	0.89	0.72	0.87	0.92
				At the bottom	1.0	1.0	1.0	1.0	1.0	1.0
3	Kegaili Channel 1978 PC 33	3.03	1.42	POV	0.35	0.5	0.78	0.42	0.57	0.8
				0.2	0.46	0.65	0.87	0.49	0.63	0.84
				0.6	0.62	0.80	0.82	0.70	0.79	0.91
				0.8	0.94	0.89	0.9	0.83	0.89	0.95
				At the bottom	1.0	1.0	1.0	1.0	1.0	1.0

Table 5.23 Towards the calculation of average vertical turbidity

Sampling location	No.	N, m	V, m/s	ω, m/s	S_0, kg/m^3	α_2	$\alpha_2 H$	$e^{-\alpha_2 H}$	$1 - e^{-\alpha_2 H}$	$S_{calculated}$, kg/m^3	$S_{measured}$, kg/m^3
1	2	3	4	5	6	7	8	9	10	Eleven	12
	1	2.58	0.87	0.00008	3.75	0.22	0.57	0.665	0.434	2.85	2.53
	2	2.96	1.1	0.00031	3.26	0.317	0.94	0.391	0.608	2.11	2.06
	3	3.48	1.34	0.00195	4.74	0.497	1.73	0.177	0.82	2.225	2.58
	4	3.68	1.33	0.00195	4.71	0.5	1.84	0.158	0.84	2.15	2.19
	5	3.74	1.01	0.00125	6.72	0.79	2.96	0.05	0.948	2.15	2.66
	6	3.3	0.79	0.00003	3.55	0.282	0.931	0.393	0.606	2.31	2.09
	7	3	1.13	0.0007	6.26	0.55	1.34	0.26	0.734	3.45	3.4
	8	2.72	1.43	0.00195	6.67	0.42	1.14	0.317	0.682	3.99	3.57
	9	1.74	0.85	0.00008	2.83	0.287	0.499	0.606	0.393	2.22	2.12
	10	2.38	1	0.00008	3.29	0.15	0.37	0.69	0.309	2.75	2.22
	Eleven	3.6	1.33	0.00008	4.65	0.395	1.42	0.24	0.758	2.48	2.23
	12	3.15	1.22	0.00008	4.55	0.093	0.294	0.74	0.255	3.95	3.7
	13	1.4	1.09	0.0007	5.46	0.49	0.688	0.5026	0.197	3.94	3.78
	14	1.85	1.18	0.00008	3.64	0.102	0.188	0.82	0.17	3.32	3.12

References

1. Abal'yants, S.Kh.: Stable and Transient Regimes in Artificial Channels, 240 p. Gidrometeoizdat, Leningrad (1981)
2. Alexandrov, V.L.: Technical Hydromechanics, 431 p. M.-L.OGIZ (1947)
3. Altunin, V.S.: Reclamation Channels in Earthen Channels, 256 p. Kolos, Moscow (1979)
4. Antsiferov, A.M., Debolsky, V.K.: Distribution of suspended matter concentration in a stationary flow over an eroded bottom. Water Resour. **24**(3), 270–276 (1997)
5. Arifjanov, A.M.: Results of studies of the sediment regime on the Karakum Canal. Abstracts of reports of the Republican Scientific and Technical Conference on Improving the Technical Operation of Canals of Irrigation Systems, pp. 61–62. Tashkent (1984)
6. Arifjanov, A.M.: On the speed of movement of a suspended flow. In the International Collection. Current Problems of Mechanics, pp. 41–45 (1995)
7. Arifjanov, A.M.: Toward the calculation of the hydraulic size of suspended sediment particles. Problems of Mechanics, vol. 5, pp. 88–90 (1995)
8. Arifjanov A.M.: Transporting capacity of flow in open channels. Problems of Mechanics, vol. 4, pp. 31–34 (1966)
9. Rifzhanov, A., Amiev, L.N.S., Akhmedov, I.G.: Irrigation value of river sediments. Current Probl. Nat. Sci. **06**(53) (2013)
10. Arifjanov, A.M., Khazratov, A.N.: Kinematics of irrigation canals. "Uzbekiston qishlok huzhaligi" magazines, No. 5-son, 24 b (2007)
11. Arifjanov, A.M., Khazratov, A.N., Usanov, M.: Transporting capacity and critical speeds of irrigation canals in an earthen channel. Republic of Ilmiy-Amalii Anzhumani Materiallari, 25–26 Nov, Tashkent, pp. 41–42 (2009)
12. Arifjanov, A.M.: Methods for calculating the distribution of sediment particles in channels of variable cross-section. Hydraul. Eng. Constr. **4**, 50–54 (2004)
13. Arifjanov, A.M.: Distribution of suspended sediment in a stationary flow. Water Resour. **28**(2), 185–187 (2011)
14. Arifjanov, A.M., Usanov, M.N.: Kanallarda notekis harakatning hususiyatlari. Tashkent, "Agro Ilm" magazine, pp. 41–42 (2010)
15. Arifjanov, A.M., Fatkhullaev, A.M.: On the stability of irrigation canals in earthen channels. In: International Scientific and Practical Conference "Problems of Development of Land Reclamation and Water Management and Ways to Solve Them" 11–14 Apr, Moscow, 2011
16. Arifjanov, A.M., Fatkhullaev, A.M.: Hydraulic calculation of irrigation channels in the earthen channel. In: Kishlok va suv hўzhaligi muammolari: Tez, Dokl. Scientific Conf. 3–6 May 2004, Tashkent, pp. 66–67
17. Akmalov, Sh.: Tindirgichdagi tinish zharayoniga hydraulic yiriklikning tasiri. Toshkent, "Agro ilm" J. 54–55 (2012)
18. Rifzhanov, A., Amiev, L.N.S., Akhmedov, I.G.: Irrigation value of river sediments. Curr. Probl. Nat. Sci. **06**(53) (2013)
19. Arifjanov, A.M., Khazratov, A.N.: Kinematics of irrigation canals. "Uzbekiston qishlok huzhaligi" magazines, No. 5-son, 24 (2007)

© The Editor(s) (if applicable) and The Author(s), under exclusive license to Springer Nature Singapore Pte Ltd. 2026

A. M. Arifjanov et al., *Dynamics of Suspended Flow in Open Channels*,
https://doi.org/10.1007/978-981-96-8781-7

20. Arifjanov, A.M.: Methods for calculating the distribution of sediment particles in channels of variable cross-section. Hydraul. Eng. **4**, 50–54 (2004)
21. Arifjanov, A.M.: Distribution of suspended sediment in a stationary flow. Water Resour. **28**(2), 185–187 (2011)
22. Arifjanov, A.M., Usanov, M.N.: Kanallarda notekis harakatning hususiyatlari. Tashkent, "Agro Ilm" Magazine, 41–42 (2010)
23. Arkhangelsky, B.V.: Experimental study of the accuracy of hydraulic particle size scales. Izv. NIIG, vol. 15, pp. 157–184 (1935)
24. Asaulenko, I.A., et al.: Theory and Applied Aspects of hydrotransportation of Solid Materials, 365 p. Naukova Duma, Kiev (1981)
25. Balakaev, B.K.: Karakum Canal, p. 194. YLYM, Ashgabat (1979)
26. Barenblatt, G.I.: On the movement of suspended particles in a turbulent flow. Appl. Math. Mech. **17**(3), 261–274 (1958)
27. Barenblatt, G.I.: Similarity, self-similarity. Intermediate Asymptotics, 255 p. Gidrometeoizdat, Leningrad (1982)
28. Baryshnikov, N.B., Popov, I.V.: Dynamics of Channel Flows and Channel Processes, 531 p. Gidrometeoizdat, Leningrad (1988)
29. Bernatskaya, N.V.: Distribution of sediment particle sizes along the depth of a suspended flow: Author's abstract, 17 p. Diss. Ph.D. tech., Nauka, Moscow (1984)
30. Borovkov, V.S.: Channel Processes and Dynamics of River Flows in Urban Areas, 286 p. Gidrometeoizdat, Leningrad (1989)
31. Buevich, Yu.A.: On the statistical mechanics of particles suspended in a gas flow. Appl. Math. Mech. **32**(1), 95–105 (1968)
32. Buevich, Yu.A.: Approximate statistical theory of a suspended layer. Appl. Math. Mech. **6**, 101–108 (1969)
33. Velikanov, M.A.: Dynamics of Channel Flows, vol. 2, 323 p. Gostekhizdat, Moscow (1955)
34. Velikanov, M.A.: Ruslovy Process, 396 p. Fizmatgiz, Moscow (1958)
35. Vinogradova, V.I.: Experimental study of velocity fields and concentrations of uniform suspension-carrying flows, 15 p. Dis. Ph.D. tech. Sciences, Moscow (1962)
36. Voynich-Syanozhentsky, T.G.: The problem of stability of flow in a real liquid in channels of finite depth. Izv. G NISGEI, vol. 18, pp. 20–39 (1969)
37. Girgidov, A.D.: Characteristics of turbulent mass transfer in open channels. J. Hydraul. Eng. Moscow **5**, 55–59 (2007)
38. Glovatsky, O.Ya., Adzhi-Veli, Y.Kh.: Some provisions of the energy approach to the theory of suspended flow. News of the Academy of Sciences of the UzSSR, vol. 6, pp. 55–60 (1986)
39. Goncharov, V.N.: Dynamics of Channel Flows, 373 p. Gidrometeoizdat, Leningrad (1962)
40. Gorbis, Z.V.: Heat Transfer and Hydromechanics of Dispersed Through Flows, 423 p. Energy, Moscow (1970)
41. Garnet, N.L.: Movement of a free solid particle in a turbulent fluid flow. Izvestia VNIIG, vol. 65, pp. 65–75 (1960)
42. Grishanin, K.V.: Hydraulic Resistance of Natural Channels, 182 p. St. Petersburg (1992)
43. Grishanin, K.V.: Discharge of channel sediments in rivers with a sandy bottom. Channel Processes and Methods of Their Modeling, pp. 8–14 (1977)
44. Grishanin, K.V.: Stability of River Beds and Canals, 144 p. Gidrometeoizdat, Leningrad (1974)
45. Grishin, N.N.: Mechanics of Bottom Sediments, 160 p. Nauka, Moscow (1982)
46. Gupalo, Yu.P., Cherepanov, G.P.: On the transition of a layer of solid particles into a suspended state. Izvestia of the USSR Academy of Sciences. Mechanics of Liquid and Gas, **1**, 105–109 (1968)
47. Arifjanov, A.M., Usanov, M.N., Khazratov, A.N.: Guvokhnoma No. DGU 02389 Oqimning notekis harakatida chўkindilar taksimoti
48. Arifjanov, A.M., Fatkhullaev, A.M., Usanov, M.N.: Guvokhnoma No. DGU 02370 Notekis harakatda okim parameterlarining hydraulics hisobi

49. Debolsky, V.K., et al. Dynamics of Channel Flows and Lithodynamics of the Coastal Zone of the Sea, 301 p. Nauka, Moscow (1994)
50. Debolsky, V.K., Dolgopolova, E.N., Zamai, O.A., Orlov, A.: Statistical description of turbulent movement in rivers. Water Resour. **4**, 57–67 (1986)
51. Debolsky, V.K., Kotkov, V.M., Stepanova, N.I.: Patterns of river sediment transport/Abstracts Dokl. V All-Union. Hydrologist. Congress, 1986, pp. 101–103
52. Dementyev, M.A., Pechenkin, M.V.: Suspended concentration fields and kinematics of suspended flows. Izvestia VNIIG, vol. 84, pp. 33–45 (1967)
53. Demidovich, B.P., Maron, I.A., Shuvalova, E.Z.: Numerical Methods of Analysis, 366 p. Nauka, Moscow (1967)
54. Denisov, Yu.M., Borovikova, L.N.: Relation of the movement of sediments (aerosols) using a generalized kinetic equation. Dynamics and Thermals of Rivers, Reservoirs, Internal and Marginal Seas, pp. 228–230. Moscow (1994)
55. Dunin, A.K.: General differential equations of two-phase flows. Izvestia Sib.o td. AN USSR **10**, 42–48 (1961)
56. Dyunin, A.K., Borshchevsky, Yu.T., Yakovlev, N.A.: Basic Mechanics of Multi-component Flows, 76 p. Publishing House of the Siberian Branch of the USSR Academy of Sciences, Novosibirsk (1965)
57. Egizarov, I.V.: Transporting capacity of open flows. Izvestia of the USSR Academy of Sciences. Dept. those. Sci. **2**, 93–108 (1966)
58. Elmanova, V.I.: Study of the rate of constrained sedimentation of fine particles in monopolydisperse systems, 27 p. Diss. Candidate of Technical Sciences, Moscow (1976)
59. Efremov, V.I., Rusin, V.A.: Generalized dependence for the hydraulic size of particles of various shapes. Meteorol. Hydrol. **4**, 93–97 (1976)
60. Zheleznyakov, G.V.: Capacity of Canal and River Beds, 311 p. Gidrometeoizdat, Leningrad (1981)
61. Zheleznyakov, G.V., Negovskaya, T.A., Ovcharov, E.E.: Hydrology, Hydrometry and Flow Regulation, 432 p. Kolos, Moscow (1984)
62. Zheleznyakov, G.V., Talmaza, V.F.: Dependence of velocity profile parameters on hydraulic resistance. Hydrotechnical Construction, vol. 8, pp. 33–35 (1973)
63. Zamarin, N.A.: Transporting Capacity and Permissible flow Velocities in Channels, 82 p. Gotransizdat, Moscow (1951)
64. Zegzhda. A.P.: Fall of sand and gravel grains in stagnant water. Izv. NIIG, vol. 12, pp. 30–54 (1934)
65. Ibad-zade, Yu.A.: Movement of Sediments in Open Channels. Stroyizdat, Moscow (1983)
66. Ibad-zade, Yu.A.: Alluvium Regime of Rivers, 325 p. Stroyizdat, Moscow (1989)
67. Ibad-zade, Yu.A.: Transporting Water in Open Channels, 273 p. Stroyizdat, Moscow (1983)
68. Irmukhamedov, H.A.: Vertical distribution of turbidity in the area of Deigisha river A Mudarya. Rep. Acad. Sci. UzSSR **2**, 71–73 (1980)
69. Ismagilov, Kh.A.: Protection of Agricultural Lands from Mudflows, 64 p. Publishing House "Mekhnat", Tashkent (1988)
70. Kaveshnikov, N.T.: Methods for calculating sediment deposition and flow turbidity parameters in settling tanks. Hydrotech. Constr. **10**, 26–30 (1997)
71. Karapetyan, M.R., Arifjanov, A.M.: Hydraulic calculation of constraint and overlap when performing channel regulation work. Probl. Mech. **1**, 51–56 (1997)
72. Karasev, I.F.: Channel Processes During Runoff Transfer, 288 p. Gidrometeoizdat, Moscow (1975)
73. Karaushev, A.V.: Theory and Methods for Calculating River Sediments, 270 p. Gidrometeoizdat, Leningrad (1977)
74. Klyukanova, I.V.: Suspended Sediments of the Amu Darya and Their Irrigation Significance, 112 p. Science, Moscow (1971)
75. Kolmogorov, A.N.: On a new version of the gravitational theory of the movement of suspended sediments M.A. Velikanova/West M. State University. Ser. Fiz. from -Mat. and Eats. Sciences **3**, 41–46 (1954)

76. Kondratyev, N.E., Popov, I.V., Snishchenko, B.F.: Fundamentals of the Hydromorphological Theory of Channel Process, 272 p. Gidrometeoizdat, Leningrad (1982)
77. Kosyan, S.D., Antsiferov, S.M., Efremov, S.A.: On the applicability of diffusion theory to calculations of sediment distribution in an open flow. Meteorol. Hydrol. **1**, 79–67 (1976)
78. Krill, S.I., Berman, V.P.: Distribution of the concentration of solid particles along the depth of a highly concentrated suspended flow. Hydromechanics **34**, 64–70 (1976)
79. Krill, S.N.: Pressure Suspension-Carrying Flows, 158 p. Naukova Dumka, Kyiv (1990)
80. Krupnik, M.Ya.: Distribution of single-grain sediments along the depth of a turbulent flow. In: Proceedings of Coordination Meetings on Hydraulic Engineering, vol. 36, pp. 58–64 (1967)
81. Krylov, A.N.: Lecture on Approximate Computing, 400 p. Gostekhizdat, Moscow (1954)
82. Kuzmin, I.A.: Forecasting of channel processes and sediment regime at water intake hydraulic structures. Proceedings of Hydroproject, pp. 154–169 (1973)
83. Lapshenkov, V.S.: Forecasting Channel Deformations in the Tailwaters of River Hydraulic Structures, 240 p. Gidrometizdat, Leningrad (1979)
84. Latipov, K.Sh.: To the determination of the coefficient of hydraulic friction. Dokl. Acad. Sci. UzSSR **8**, 16–18 (1982)
85. Latipov, K.Sh.: On internal friction stresses in a liquid. Izvestia Acad. Sci. UzSSR Ser. Tech. Sci. **6**, 43–47 (1980)
86. Latipov, K.Sh.: Theory of Motion of Multiphase Media and Its Applications to Some Technical Issues: Abstract of Thesis, 40 p. Diss. Ph.D. tech. Nauka, Tashkent (1970)
87. Latipov, K.Sh., Arifjanov, A.M.: Issues of the Movement of Suspended flow in Open Channels, 110 p. Publishing House "Mekhnat", Tashkent (1994)
88. Latipov, K.Sh., Arifjanov, A.M.: To determine the nature of the distribution of suspended sediment particles along the depth of the flow. Izvestia Acad. Sci. UzSSR. Ser. Tex. Sci. **3**, 50–52 (1984)
89. Latipov, K.Sh., Arifjanov, A.M.: Towards the calculation of turbidity distribution along the length of a channel of variable cross-section. Reports of the Academy of Sciences of the UzSSR, No. 4, pp. 23–25 (1987)
90. Latipov, K.Sh., Arifjanov, A.M.: On the model of the movement of suspension-carrying flow in channels. Mech. Probl. **6**, 51–54 (1996)
91. Latipov, K.Sh., Arifjanov, A.M., Fatkhullaev, A.M., Ilkhomov, Kh.: Turbulent flows in pressure systems. J. Mech. Probl. Tashkent **2**, 33–38 (2005)
92. Latipov, K.Sh., Arifjanov, A.M.: To determine the stable cross-section of an irrigation canal. J. Probl. Mech. Tashkent **2**, 32–35 (1997)
93. Latipov, K.Sh., Fatkhullaev, A.M.: Turbulent tangential stresses. In: Development of water Management and Land Reclamation of the Republic of Uzbekistan During the Transition to a Market Economy: Res. Scientific-Practical Conference, Tashkent, pp. 84–85 (2006)
94. Latipov, K.Sh., Arifjanov, A.M., Fatkhullaev, A.M.: Turbulent flows of liquid flow in irrigation systems. In: International Scientific and Practical Conference "Problems of Development of Land reclamation and Water Management and Ways to Solve Them", 11–14 Apr, Moscow (2011)
95. Latipov, K.Sh., Umarov, A.I.: On the coefficient of interaction of multiphase media. Questions of Mechanics, vol. 7, pp. 57–60 (1969)
96. Latipov, K.Sh., Shoyusupov, M.: About Channel Flows with Variable Flow along the Path, 192 p. Fan, Tashkent (1979)
97. Levi, I.I.: Dynamics of Channel Flows, 252 p. Gosenergioizdat, Moscow (1957)
98. Levi, I.I., Kulesh, N.P.: Transporting ability of a flow saturated with fine suspended matter. Izvestia VNIIG **66**, 3–20 (1960)
99. Loytsyansky, L.G.: Mechanics of Liquid and Gas, 847 p. Nauka, Moscow (1973)
100. Lyapin, A.N.: Hydraulic resistance during the transition from laminar to turbulent flow. Proceedings of the VODGEO Institute, pp. 21–27 (1990)
101. Lyapin, A.N., Alexandrova, A.K.: Calculation of deep deformations of natural rivers. In: Proceedings of the State Hydrological Institute, vol. 263, pp. 78–85 (1980)

102. Lather, V.M., Prudovsky, A.M.: Hydraulic Modeling, 340 p. Energoatomizdat, Moscow (1984)
103. Magomedova, A.V.: Forecast of erosion processes and sediment transport. Author's abstract. diss. Ph.D. tech., 41 p. Nauka, Moscow (1993)
104. Makkaveev, V.M.: On the theory of turbulence and sediment movement. Proceedings of the State Hydrological Institute, vol. 160, pp. 54–87 (1963)
105. Makkaveev, V.M.: On the theory of turbulent regime and sediment suspension. Proceedings of the State Hydrological Institute, vol. 32, pp. 5–26 (1931)
106. Makkaveev, V.M.: Movement of solid particles in turbulent fluid flows. In: Proceedings of Coordination Meetings on Hydraulic Engineering, vol. 36, pp. 11–18 (1967)
107. Makkaveev, N.I., Chalov ,R.S.: Channel Processes, 260 p. Moscow State University Publishing House, Moscow (1986)
108. Mamedov, M.A.: Experimental studies of averaged local characteristics of suspension flows: author's abstract. diss, 10 p. Ph.D. tech. Nauka, Moscow (1967)
109. Mikhailova, N.A.: Transport of Solid Particles by Turbulent Water Flows, 234 p. Gidrometeoizdat, Leningrad (1966)
110. Mikhalev, M.A.: Differential equations and integral relations of the movement of a viscous heavy fluid in an open channel. Channel Processes and Methods of Their Modeling, pp. 30–38 (1977)
111. Mukhamedov, A.M., Zhuraev, T.: Distribution of averaged longitudinal velocities along the depth of the flow of the Mudarya River. Reports of the All-Union Academy of Agricultural Sciences No. 5, pp. 24–27 (1973)
112. Mukhamedov, A.M., Ismagilov, H.A.: Turbidity of the flow of temporary watercourses. Reports of VASKhNIL, no. 1, pp. 36–38 (1981)
113. Myasnikov, V.P. On collisions of solid particles in liquid. Rep. USSR Acad.. **185**(6), 1258–1261 (1969)
114. Natishvili, O.G.: On some features of the movement of suspended flow in open channels. Proceedings of Coordination Meetings on Hydraulic Engineering, vol. 37, pp. 32–41 (1967)
115. Natishvili, O.G., Gogeshidze, I.M.: On the basic equations of hydrodynamics describing the movement of a suspension-carrying flow. Proceedings of GruzNIIGiM, vol. 29, pp. 50–58 (1972)
116. Nigmatullin, R.I.: Fundamentals of Mechanics of Heterogeneous Media, 336 p. Nauka, Moscow, 1978
117. Nikitin, I.K.: Turbulent Channel Flow and Processes in the Bottom Region, 142 p. Kyiv: Publishing House of the Ukrainian SSR Academy of Sciences (1963)
118. Obrazovsky, A.S.: On the question of the mechanism of sediment suspension by a uniform turbulent flow. In: Proceedings of VNIIVodGEO, vol. 5, pp. 63–78 (1957)
119. Pechenkin, M.V.: Changes in the turbulent characteristics of a flow by solid particles suspended in it. In: Proceedings of Coordination Meetings on Hydraulic Engineering, vol. 57, pp. 63–67
120. Pechenkin, M.V.: Measuring the kinematic characteristics of turbulent flows using the optical method. River Hydraulics and Channel Processes Part 1, pp. 118–124 (1976)
121. Proskuryakov, A.K.: On the question of two theories of suspended sediment transfer. Problems of Channel Processes, pp. 82–87 (1953)
122. Pyrkin, Yu.G., Silaev, M.A.: The influence of turbulence on the rate of sedimentation of Stokes particles in a near-bottom suspension flow. Dynamics and Thermals of Rivers, Reservoirs, Internal and Marginal Seas, pp. 310–311 (1994)
123. Rabkova, E.K.: Methodology for calculating channel reformations for a flow of high turbidity. River Hydraul. Hydraul. Eng. **15**(7), 91–104 (1973)
124. Rabkova, E.K.: Design and Calculation of Irrigation Canals in Earthen Channels, 271 p. U.D.N., Moscow (1990)
125. Razumikhina, K.V.: Experience in calculating sediment flow. Proc. State Hydrol. Inst. 191, pp. 100–110 (1972)
126. Rouse, H.: Fluid Mechanics, 389 p. Mir, Moscow (1967)

127. Rakhmatullin, H.A.: Fundamentals of hydrodynamics of interpenetrating motions. Appl. Math. Mech. **20**(2), 184–195 (1956)
128. Reynolds, A.: Turbulent Flows in Engineering Applications, 408 p. Energy, Moscow (1979)
129. Rzhanitsyn, N.A.: Morphometric and Hydrological Patterns of the Structure of the River Network, 238 p. Gidrometeoizdat, Leningrad (1960)
130. Rzhanitsyn, N.A.: Channel-Forming Processes of Rivers, 285 p. Gidrometeoizdat, Leningrad (1985)
131. Romanovsky, V.V.: The influence of the shape of large sediment particles and their orientation in the flow on hydrodynamic resistance. Proc. GGI **175**, 108–118 (1969)
132. Romanovsky, V.V.: Experimental study of hydraulic sediment size. Proc. State Hydrol. Inst. **191**, 111–136 (1972)
133. Rossinsky, K.I., Arbulieva, K.M.: Flow around bodies in a turbulent flow near a solid surface. Water Resour. **6**, 156–167 (1977)
134. Rossinsky, K.I., Debolsky, V.K.: River Sediments, 215 p. Nauka, Moscow (1960)
135. Samolyubov, B.I., Sluev, M.V.: Mathematical modeling of the velocity field of a suspension current. Dynamics and Thermals of Rivers, Reservoirs, Internal and Marginal Seas, pp. 318–320. Moscow (1994)
136. Sedov, L.I.: Methods of Similarity and Dimension in Mechanics, 440 p. Science, Moscow (1972)
137. Silin, N.A., Vitoshkin, Yu.K., Karasik, V.M., Ocheretko, V.F.: Hydrotransport, 160 p. Naukova Dumka, Kyiv (1971)
138. Silin, N.A., Vitoshkin, Yu.K., Kerasik, V.M.: Distribution of solid particle concentration along the depth of a suspended flow. In: Proceedings of Coordination Meetings on Hydraulic Engineering, vol. 57, pp. 34–39 (1971)
139. Silin, N.A., Pishchenko, I.A., Ocheretko, V.F.: Experimental study of the distribution of average velocities, consistencies and particle sizes across a pipeline cross-section during the movement of two-component flows. Hydrotechnics and Hydromechanics, pp. 62–69. Kyiv (1964)
140. Sinelshchikov, V.S.: On the diffusion coefficient of suspended particles in turbulent flow. In: Proceedings of Coordination Meetings on Hydraulic Engineering, Issue 36, pp. 19–25 (1967)
141. Slezkin, N.A.: Dynamics of a Viscous Incompressible Fluid, 519 p. Gostekhizdat, Moscow (1955)
142. Sou, S.: Hydrodynamics of Multiphase Systems, 536 p. Mir, Moscow (1971)
143. Syromyatnikov, N.I.: Statistical Theory of Suspended Layer Formation. Dokl. ANNA USSR TEBR **93**(3), 421–424
144. Talmaza, V., Kroshkin, A.N.: Hydromorphometric Characteristics of Mountain Rivers, 204 p. Kyrgyzstan, Frunze (1968)
145. Tikhonov, A.N., Samarsky, A.A.: Equations of Mathematical Physics, 636 p. Nauka, Moscow (1977)
146. Tuzov, V.B., Karapetyan, M.R.: Study of sediments in the lower rivers of the Mudarya. Works of SredneazNIIrrigatsiya, vol. 121, pp. 37–42 (1976)
147. Umarov, A.I.: Hydrodynamics of Multiphase Mixtures in Pipes and Porous Media. Diss. Ph.D. tech, 35 p. Science, Leningrad (1986)
148. Faizullaeva, D.F.: Laminar Movement of Multiphase Media in Pipelines, 220 p. Science, Tashkent (1966)
149. Faizullaev, D.F., Umarov, A.I., Shakirov, A.I.: Hydrodynamics of One and Two-Phase Media and Its Practical Application, 168 p. Fan, Tashkent (1986)
150. Fatkhullaev, A.M., Yusupov, M., Nizamutdinov, D.R.: The influence of the concentration of suspended sediment particles on the distribution of average flow velocity. Collection of Scientific Papers, SANIIRI Tashkent, pp. 447–450 (2005)
151. Fatkhullaev, A.M., Aknazarov, O.: On the cross-sectional shape of stable earthen channels. SANIIRI on the way to integrated water resources management. Collection of Scientific Works SANIIRI, pp. 161–165 (2010)

152. Fatkhullaev, A.M., Khazratov, A.N.: Limit regimes of irrigation canals in earthen beds. Agro Illm. **3**(15), S-41–42 (2010)
153. Fidman. B.A.: On the distribution of solid particles in a vertical fluid flow. Izvestia USSR Acad. Sci. Dept. Tech. Sci. **3**, 175–176 (1963)
154. Fidman, B.A.: Main results of experimental study of the structure of turbulent flows. Problems of Riverbed Processes, pp. 138–150 (1953)
155. Fidman, B.A.: Turbulence of Water Flows, 240 p. Gidrometeoizdat, Leningrad (1991)
156. Frankl, F.I.: On the theory of the movement of suspended sediments. Dokl. USSR Acad. Sci. **92**(2), 247–250 (1953)
157. Frankl, F.I.: Experience of a semi-empirical theory of the movement of suspended sediment in an uneven flow. Dokl. Acad. Sci. USSR **102**(6), 250–259 (1955)
158. Frankl, F.I.: Energy equations for the movement of fluid with suspended sediment. Dokl. Acad. Sci. USSR **102**(5), 903–906 (1955)
159. Fuks, N.A.: Mechanics of Aerosols. Publishing House of the USSR Academy of Sciences, Moscow (1955)
160. Hunt D.N. Dynamics of an Incompressible Fluid, 184 p. Mir, Moscow (1976)
161. Happel, J., Brenner, G.: Hydrodynamics at Low Reynolds Numbers, 630 p. Mir, Moscow (1976)
162. Khachatryan, A.G.: Settlement Basins on Irrigation Systems, 342 p. Selkhozgiz, Moscow (1957)
163. Khachatryann, A.G., Shapiro, H.Sh., Sharova, E.N.: Siltation and Leaching of Irrigation Sedimentation Basins and Reservoirs, 236 p. Kolos, Moscow (1966)
164. Khmalidze, G.N.: On the issue of the relationship between the flow rates of transported and suspended sediments. Channel Processes **10**, 164–171 (1973)
165. Tsvetikova, N.F.: Methodology for calculating sediment deposition in a turbulent flow. Bed Processes, pp. 124–132. Publishing House of the USSR Academy of Sciences, Moscow (1958)
166. Tsvetkova, N.A.: Sediment regime of the Mudarya river. Tashkent—Issues of Hydraulic Engineering, pp. 5–86 (1963)
167. Chebotarev, A.I.: Hydrological Dictionary, 308 p. Gidrometeoizdat, Leningrad (1978)
168. Chugaev, R.R.: Hydraulics (Technical Fluid Mechanics), 600 p. Energy, Leningrad (1975)
169. Shamov, G.N.: River Sediments, 378 p. Gidrometeoizdat, Leningrad (1959)
170. Shapiro, H.Sh.: Regulation of Solid Runoff During Water Intake into Irrigation Systems, 271 p. Kolos, Moscow (1983)
171. Sharenkov, I.A.: Applied Planning Problems of Quiet Flow Hydraulics, 240 p. Energy, Moscow (1978)
172. Shashin, V.M.: Hydromechanics, 383 p. Higher School, Moscow (1990)
173. Shebalin, O.D.: Molecular Physics, 175 p. Higher School, Moscow (1978)
174. Schlichting, G.: Boundary Layer Theory, 712 p. Nauka, Moscow (1974)
175. Shterenlikht, D.V.: Hydraulics, 635 p. Energoatomizdat, Moscow (2001)
176. Shcheglova, O.P.: Formation of Suspended Sediment Runoff and Washout of the Mountainous Part of Central Asia, 237 p. Gidrometeoizdat, Leningrad (1972)
177. Yufin, A.N.: Sediment Movement and Hydraulic Transport, 200 p. Gosenergoizdat, Leningrad (1963)
178. Ackers, P., White, W.R.: Sediment transport: new approach and analysis. Proc. ASCE Hydr. Div. **99**(11), 2041–2060 (1973)
179. Alder, G.R., Simons, D.B.: Fall velocity of irregular shaped particles. Proc. ASCE (J. Hydr. Div.) **94**(HV3), 721–737 (1968). Paper 5949
180. Bogardi, I.: Sediment Transport in Alluvial Streams, p. 826. Akademia Kiado, Budapest (1974)
181. Chepil, W.S.: The use of evenly spaces hemispheres to evaluate aerodynamic forces on a soil surface. Trans. Am. Geophys. Union **39**(3), 397–404 (1958)
182. Chien, N.: The present status of research on sediment transport. Proc. Amer. Soc. Civ. Eng. **80**, 70 (1954)

183. Coles, D.: The law of the wake in the turbulent Boundary layer. J. Fluid Mech. **4**(pt. 1), 191–226 (1956)
184. Einstein, X.A., Abdel-Aalf, F.M.: Einstein bed load function at high sediment rates. J. Hydraul. Div. Proc. Am. Soc. Civ. Eng. **98**(1) (1972)
185. Engelund, F.: Instability of erodible beds. J. Fluid Mech. **42**(2), 225–244 (1970)
186. Entenuer, G.A.: Ein Mechanismus des Einseters on Geschielebewegungeen in Stromungln-Forsch. Ind. Bd. 35 (1969)
187. Fage, A.: Fluid flow in rough piper. Aeronautical Research Committee Reports and Memoranda, No. 1585 (1933)
188. Fischer, H.B.: Longitudinal dispersion and turbulent mixing in open-channel flow. Annu. Rev. Fluid Mech. **5** (1973)
189. Fisher, H.B.: The mechanics of dispersion in natural streams. J. Hydraul. Div. ASCE **93**(6) (1967)
190. Fukustima, Y., Parker, G., Pantin, H.: Prediction of ignitive turbidity currents in Scripps Submarine Canyon. Mar. Geol. (1985)
191. Graf, W.H.: Hydraulics of sediment transport, p. 513. McGraw Hill, New. York (1971)
192. Ismail, H.M.: Turbulent mechanism and suspended sediment in closed channels. Proc. ASCE **77**(56), 26–52
193. Hino, M.: Turbulent flow with suspended particles. J. Hydraul. Div. Proc. Amer. Soc. Civ. Eng. **89**, 513 (1963)
194. Hunt, I.N.: On turbulent transport of a heterogeneous sediment. Q. J. Mech. Appl. Math. **22**, 235–246 (1969)
195. Insjumoto, T., Mori, A., Okabe, T., Ohmoto, T.: Non-equilibrium sediment transport: a generalized model. J. Hydrosci. Hudronl. Eng. **7**(7), 1–25 (1990)
196. Jobson, H.E., Sayre, W.W.: Vertical transfer in an open channel. How. J. Hydraul. Div. Proc. Amer. Soc. Civ. Eng. **3**(4), 703–724 (1970)
197. Kironoto, B., Graf, W.H.: Decelerating flow in a gravel bed flume. Rapp. Annu. Lab. d. Hydroul. EPFL, Lausanne, pp. 1–14 (1989)
198. Lazzari, F.: Cofficienti di resistensa spaerimentali verso picooli valori delnumero di Reynolds per ellipssoidi rotondi schiacciati. Energ. Elec. Milano **53**(6), 293–313 (1976)
199. McQuivery, R.S.: Summary of turbulence data from rivers, conveyance channels and laboratory flumes. US Geol. Surv. Prof. Paper 802 B (1973)
200. Molerus, O.: Zur Beschreibind fest stoffbeladerner stromunden. Chem. Ind. Technol. Bd **49**(12) (1977)
201. Moreland, C.: Viscosity of suspensions of geol. in mineral oil. Can. J. Chem. Eng. **41**(1), 24–30 (1963)
202. Saffman, P.G.: On the boundary condition at surface of a porous medium. Stud. Appl. Math. **50**(2), 93–101 (1971)
203. Saffman, P.G.: On the spectrum and decay of random two-dimensional vorticity distributions at large Reynolds number. Stud. Appl. Math. **50**(4), 377–383 (1971)
204. Siacey, M., Bowen, A.: The vertical structure of density and turbidity currents. J. Geophys. Rec. **93**, 3528 (1988)
205. Soo, S.L., Tien, C.L.: Effect of the wall on two-phase turbulent motion. Trans. ASME Ser. E **27**(1) (1960)
206. Taylor, G.I.: The dispersion of matter in turbulent flow through a pipe. Proc. Roy. Soc. A **223**(1155) (1954)
207. Vanoni, V.A.: Resistance properties of sediment streams. Trans. ASCE **125**(Part 1), 1140–1173 (1960)
208. Vanoni, V.A.: Transport of suspended sediment by water. Proc. ASCE **70**(6), 793–828 (1944)
209. Vanoni, V.A., Nomicos, G.N.: Resistance properties of sediment Laden streams. J. Hydraul. Div. ASCE **85**(H45), 77–107 (1959)
210. Yalin, M.S.: Mechanics of Sediment Transport. Pergamon Press, London (1967)
211. Yuqian, L., Ning, Q.: Erosion and transport of sediment in the Yellow river basin. Sediment Res. **16**(1), 2–38 (1986)

MIX
Papier aus verantwortungsvollen Quellen
Paper from responsible sources
FSC® C105338

If you have any concerns about our products,
you can contact us on
ProductSafety@springernature.com

In case Publisher is established outside the EU,
the EU authorized representative is:
Springer Nature Customer Service Center GmbH
Europaplatz 3, 69115 Heidelberg, Germany

Printed by Libri Plureos GmbH
in Hamburg, Germany